Fritz H. Frimmel
Frank von der Kammer
Hans-Curt Flemming

Colloidal Transport in Porous Media

Fritz H. Frimmel
Frank von der Kammer
Hans-Curt Flemming
(Editors)

Colloidal Transport in Porous Media

With 99 Figures

 Springer

PROF. DR. DR. HABIL. FRITZ H. FRIMMEL
University of Karlsruhe
Engler-Bunte-Institute
Department of Water Chemistry
Engler-Bunte-Ring 1
76131 Karlsruhe
Germany

Email: fritz.frimmel@ebi-wasser.uni-karlsruhe.de

DR. FRANK VON DER KAMMER
University of Vienna
Department of Geological Sciences
Althanstraße 14
1090 Vienna
Austria

Email: frank.kammer@univie.ac.at

PROF. DR. HANS-CURT FLEMMING
Faculty of Chemistry – Biofilm Centre
University of Duisburg-Essen
Geibelstraße 41
47057 Duisburg
Germany

Email: HansCurtFlemming@compuserve.com

Library of Congress Control Number: 2007923066

ISBN-13   978-3-540-71338-8   Springer Berlin Heidelberg New York

Springer is a part of Springer Science+Business Media
springer.com
© Springer-Verlag Berlin Heidelberg 2007

Cover design: deblik, Berlin
Typesetting: camera-ready by the editors
Production: Christine Adolph
Printing: Krips bv, Meppel
Binding: Stürtz AG, Würzburg

Printed on acid-free paper   32/2132/ca 5 4 3 2 1 0

Preface

Starting from the dimension of one or two meters which is typical for the size of human beings, we can easily discover the million fold up and down scales. However, three more digits in either direction are already a challenge for our perception, and we need sophisticated instrumental tools and analogy concepts to obtain an image of the material present. Therefore these domains have remained more unclear in many aspects of the properties and fate of the matter concerned. Particles in the smaller dimension - more exactly of the size from 1 nanometer to 1 micrometer - are normally called colloids and have recently gained a renaissance of attention of physicists, chemists, material scientists and engineers. A broad application of tailored nano particles in the daily life has been the result. Also geogenic nano and micro particles and their important role in the transport and distribution of living and non-living matter in nature have gained awareness of the scientists. In contrast to the chemo-biological importance of these micro particles stands the limited knowledge about their structure, function and fate. In aquatic sciences, they therefore have been addressed as "neglected third phase" besides the purely dissolved and gaseous matter transport mechanisms in porous media.

Reason enough for a cooperative research project on the "Colloidal Transport of Substances during the Seepage of Rain Water" (KORESI). The four institutes involved were 1. the coordinating Chair of Water Chemistry, Universität Karlsruhe (TH), 2. the Institute of Environmental Technology, Technische Universität Hamburg-Harburg, 3. the Biofilm Centre, Aquatic Microbiology, Universität Duisburg-Essen, and 4. the Institute of Water Quality Control, Technische Universität München. The work focused on the interaction of geo colloids and pollutants, the retardation and transport of colloids, the influence of biofilms and the retardation of pollutants in engineered sorption barriers. The aim of the project was the determination and assessment of the different influences on the colloidal transport of pollutants to provide recommendations for the handling of rain run-off as an essential component of ground water protection and long term water management in general.

At the end of the four years period of research, the results were presented in a workshop and discussed in the context of the international scientific community. The result was an even broader view on the manifold aspects of the occurrence and determination of colloids and of the colloid mediated transport of contaminants in soil and other porous media. Most of the contributions given are presented in this book.

I want to acknowledge the work of the authors and reviewers that was put into the manuscripts. Special thanks go to George Metreveli for handling the manuscripts and for managing all the different contacts within the cooperative project and beyond it. How could good research work without funding? On behalf of the whole group of scientists involved, I want to express our strong gratitude to the Deutsche Forschungsgemeinschaft for supporting our work. Dr. Ute Weber's great help and understanding of our needs deserves special mentioning as do the critical but helpful comments of the anonymous referees. Last but not least the publishing house put our written ideas in fine cooperation into this appealing volume.

The results of the projects shown in this book are not the only benefits of our work. There was a fine and rewarding personal interaction amongst the participants and the invited guests. Learning, discussing, searching for the best way to answer urging scientific questions were components of our way to new colloid frontiers. The senior scientists are also happy that there was plenty of scientific and social interaction with the graduate students. It seems as if the manifold results and the synergistic effect which have come out of this cooperative project reflect one of its major findings: small particles show great effects.

Fritz H. Frimmel
Project Coordinator and
Professor of Water Chemistry
Universitaet Karlsruhe

www.uni-karlsruhe.de

Contents

Transport Phenomena

1 Colloid Facilitated Transport in Natural Porous Media: Fundamental Phenomena and Modelling.

Daniel Grolimund[1], Kurt Barmettler[2], Michal Borkovec[3]

[1] Waste Management Laboratory, Nuclear Energy and Safety Department and Swiss Light Source, Paul Scherrer Institute, CH-5232 Villigen, Switzerland
[2] Institute of Biogeochemistry and Pollutant Dynamics, Soil Chemistry, Swiss Federal Institute of Technology ETH Zurich, CH-8092 Zürich, Switzerland
[3] Department of Inorganic, Analytical, and Applied Chemistry, University of Geneva, CH-1211 Geneva 4, Switzerland
e-mail: daniel.grolimund@psi.ch

1.1 Introduction

Reactive transport phenomena, in particular, the transport of contaminants, are of fundamental interest in environmental sciences. The presence of hazardous chemicals in the subsurface environment has become an important driving force to develop reactive transport models capable to predict their fate (Dagan 1989; Jury and Roth 1990; Sardin et al. 1991; Knox et al. 1993; Appelo and Postma 1996; Lichtner et al. 1996). These models represent the natural porous medium as two types of phases: (i) *immobile solid* phases and (ii) *mobile liquid* (and/or gaseous) phases. Depending on the affinity to the respective phases, chemical species distribute between the different phases and the corresponding phase boundaries. Accordingly, the transport of chemicals is dictated by partitioning of the mobile dissolved species and the stationary species adsorbed to the solid phase. Distribution into the solid phases and interfacial reactions result in a reduction of the dissolved contaminant concentrations in the liquid phase, and accordingly in a slow-down of the contaminant spreading.

Over the past years, an increasing number of investigations documented that mobile contaminants associate with solid colloidal particles dispersed in the flowing fluid phase (Champlin and Eichholz 1968; Champ et al. 1982; Vinten and Nye 1985; Buddemeier and Hunt 1988; Fauré et al.

1996; Grolimund et al. 1996; Saiers and Hornberger 1996; Roy and Dzombak 1997; de Jonge et al. 1998; Kersting et al. 1999; Grolimund and Borkovec 2006). These studies provide ample evidence that the suspended colloids do act as contaminant carriers, and represent a rapid transport pathway for highly reactive pollutants. Such an enhanced spreading of hazardous chemicals is generally referred to as "colloid facilitated transport". The recognition of suspended colloidal carriers led to the development of three-phase transport models including - as a third phase - a *mobile solid* phase (Corapcioglu and Jiang 1993; Fauré et al. 1996; Roy and Dzombak 1998; van de Weerd et al. 1998; Lenhart and Saiers 2003; Sen et al. 2004). A schematic representation of such a three-phase reactive transport system is depicted in Fig. 1.1a.

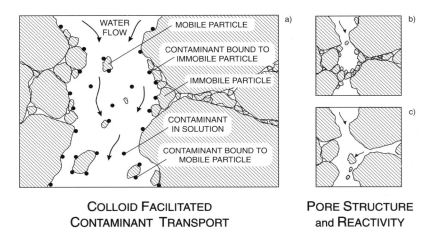

Fig. 1.1. Colloidal phenomena affecting the mobility of contaminants in porous media. a) Colloid facilitated transport. Contaminants (•) are either dissolved in the aqueous phase, or adsorbed on the surface of the solid phase. Generally, the solid phase is assumed to be immobile, but colloids can be released from the matrix, and act as mobile contaminant carriers. b) Structural damage. Breakdown of aggregates, deposition of mobile colloids or filtration of suspended aggregates cause the clogging of pores. c) Reduction in chemical reactivity of the stationary phase due to the release of colloids with a high specific surface area. Adapted from Ref. (Grolimund et al. 1996).

While discussing the potential impact of mobile colloidal particles on reactive transport, two important secondary effects should be considered as well. First, processes involving colloidal particles can potentially alter the physical structure of the porous media. *In-situ* mobilisation – at least initially – may increase the pore space, and modify its accessibility and connectivity. More importantly, however, chemical conditions favoring particle release also raise the susceptibility concerning destabilization and breakdown of aggregates within the porous media. Further, deposition of mobile colloids and filtration of suspended aggregates can cause clogging of pores (see Fig. 1.1b). Accordingly, as a result of particle release and deposition, the hydraulic properties of the subsurface porous medium can be modified. Possible consequences can be as far reaching as structural damage, including complete clogging of the porous medium.

Further, the modification of the chemical reactivity of the porous medium must be considered. A reduction in the reactivity can arise due to the mobilisation of colloidal particles. Colloids are characterized by a high specific surface area. As shown by Wu et al. (1993), a considerable fraction of the total surface area of subsurface porous media is associated with the colloidal size fraction. Consequently, along with mobilized colloids a substantial portion of the reactive surface will be eliminated from the system. A reduction in specific area is directly linked to a reduction in interfacial reactivity. This effect reduces the sorption capacity of the medium and leads to a diminishing of the retention capacity for reactive contaminants. Moreover, the spreading of a succeeding contaminant plume will be enhanced after a mobilisation event.

Due to the importance of these effects, the consideration of colloid facilitated transport, structural damage, and reduced chemical reactivity should become an essential part in the risk assessment of any subsurface contamination problem, in the development of remediation strategies, as well as in recharge and waste water management. Reliable predictions of the susceptibility of a subsurface system for scenarios, where mobile colloidal particles are of importance, require a detailed understanding of the fundamental mechanisms and the corresponding kinetics. Most relevant phenomena are (i) the generation of mobile particles, (ii) the persistence of these particles within a system, and in case of colloid facilitated transport (iii) the interactions between contaminants and mobile particles as well as between contaminants and immobile solid phase (Ryan and Elimelech 1996; Kretzschmar et al. 1999). Presently, however, even of the simplest processes of this type are hardly understood.

In the present chapter, various aspects of facilitated contaminant transport by mobile colloid particles will be addressed. The following discussion is based on a comprehensive set of laboratory-scale column experiments, which were performed to investigate relevant processes, including particle mobilisation, particle deposition, particle aggregation, and multicomponent contaminant transport (Grolimund et al. 1998; Grolimund and Borkovec 1999; Grolimund et al. 2001b; Grolimund and Borkovec 2001;

Grolimund et al. 2001a; Grolimund and Borkovec 2006). The knowledge about the structure and dynamics of these processes obtained from the individual experimental investigations was compiled and incorporated into an extended contaminant transport model. This model was used to analyze and to predict contaminant transport experiments, where *in-situ* mobilized colloidal particles have been proven to be the dominant transport pathway (Grolimund et al. 1996; Grolimund and Borkovec 2005).

1.2 Colloidal Particles

Based on their nature and composition, one can differentiate between three main classes of potentially hazardous colloidal particles: (i) intrinsic colloidal particles, (ii) carrier colloidal particles, and (iii) biocolloids (van der Lee et al. 1994; Ryan and Elimelech 1996; Kretzschmar et al. 1999). For the intrinsic colloidal particles, the contaminant represents a structural component of the colloid. Precipitates or solid solutions of inorganic contaminants are typical examples of this class. The carrier colloidal particles bind the contaminants, which are present in the moving fluid phase, to the surface of suspended organic or inorganic colloids. Viruses or bacteria are important examples of organisms in the colloidal size range representing the class of biocolloids.

Mobile colloidal particles are ubiquitous in subsurface waters, but our knowledge of their release and generation mechanisms is limited. Various sources of mobile particles have been postulated. Intrinsic colloidal particles are formed by precipitation out of supersaturated solutions. Systems with varying redox conditions are most susceptible to the formation of intrinsic colloids. Translocation from the vadose zone or recharge areas represents the major source of biocolloids. Possible sources of carrier colloids include disaggregation, dissolution of cementing mineral phases, atmospheric deposition, or artificial introduction into the system by human activities. However, dispersion phenomena resulting in the *in-situ* release of preadsorbed colloidal particles is generally considered as the most relevant source of mobile colloidal particles in subsurface systems (Ryan and Elimelech 1996; Kretzschmar et al. 1999).

Based on the relevance concerning subsurface porous media such as soils, rock and aquifers, we focus on (inorganic) carrier colloids in the following.

1.3 Fundamental Phenomena

1.3.1 *In-situ* Mobilisation of Particles

The mobilisation of preadsorbed colloids due to dispersive conditions corresponds to the most important source of mobile colloidal particles in subsurface systems. The detailed mechanisms and appropriate kinetic processes of the release phenomena within natural systems, however, are still poorly understood. Field and laboratory scale investigations document a strong coupling between the observed release rates and the chemical and physical conditions. Pronounced changes in the suspended colloid concentration have been observed after physical and chemical disturbances (Nightingale and Bianchi 1977; Gschwend and Reynolds 1987; Buddemeier and Hunt 1988; Wiesner et al. 1996).

In general, one can distinguish between physical and chemical disturbances. In advective systems, *chemical disturbances* are developing into chromatographic fronts traveling through the porous media. Changes in the concentration of inert tracers result in non-retarded fronts. Variations in ionic strength represent the most common non-retarded chemical disturbances. Typical examples include the intrusion of rainwater or the infiltration of high salt contaminant plumes. In contrast to inert tracers, reactive chemicals lead to the development of retarded fronts with a reduced spreading velocity within the porous medium (e.g., cation exchange fronts, retarded pH breakthrough, etc.). Common for both cases, the moving fronts divide the porous medium into different compartments with different chemical conditions and corresponding differences in dispersive properties.

As an illustrative example, let us consider an initially stable porous medium such as a top soil. The pore water is characterized by a moderate ionic strength. The infiltration of rainwater of lower ionic strength corresponds to a chemical disturbance of the system, and a normality front develops. As the infiltration continues, this normality front propagates through the porous medium. Downstream of the front one still observes stabilizing conditions, while the low ionic strength conditions upstream of the front favor colloid release. Consequently, the release of colloids is coupled to the migration of a non-retarded chromatographic front, which moves with the velocity of the pore water.

On the other hand, certain *physical disturbances* result in an immediate change of the release rate coefficients within the entire medium. Examples of physical disturbances include changes in flow velocity, application of external electric or magnetic fields, and – under certain circumstances – changes in temperature.

The theoretical analysis of these situations yields distinct release pattern in each case (Grolimund and Borkovec 2001). As shown in Fig. 1.2, the temporal position and the shape of the initial appearance correspond to a signature of the causing disturbance. In the case of a physical disturbance (Fig. 1.2, *left*), colloids appear immediately at the column outlet and their concentration rises gradually. The maximum concentration is reached at one pore volume. In contrast, if the mobilisation is due to a non-retarded chemical disturbance, the appearance of colloids is delayed by one pore volume. The steepness of the following raise as well as the position of the peak maximum is highly influenced by the dispersivity of the porous medium (Fig. 1.2, *middle*). In case of a retarded chemical front, the colloid concentration increases more gradually after one pore volume compared to the two previous cases. The position of the peak maximum strongly depends on the retardation of the reactive front and the dispersivity of the medium.

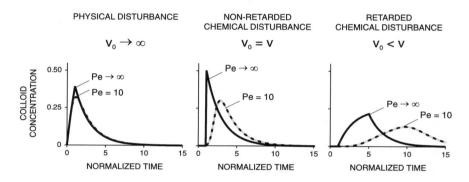

Fig. 1.2. Colloid release pattern induced by physical or chemical disturbances. *left*) Physical disturbance resulting in an immediate change in the rate constant. *middle*) Non-retarded chemical disturbance such as a normality front. *right*) Retarded chemical disturbance. Adapted from Grolimund and Borkovec (2001).

Experimental information about the general nature of the particle release process was obtained by studying mobilisation phenomena in different natural porous media by means of laboratory column experiments under well-controlled conditions. The following observations represent fundamental characteristics of the release process: (i) a pronounced non-exponential release behavior, (ii) a finite supply of colloidal particles, and (iii) a strong dependence of the observed release kinetic on chemical system parameters (Grolimund and Borkovec 1999; Grolimund et al. 2001b; Grolimund and Borkovec 2001; Grolimund and Borkovec 2006).

Fig. 1.3 depicts the outcome of a typical mobilisation experiment. The left panel shows the salt concentration and particle concentration at the column outlet in a linear representation. The corresponding double-logarithmic representation is given in the right panel. Particle release is induced by a step-wise reduction in ionic strength corresponding to a chemical disturbance (arrows in Fig. 1.3). Prior to the disturbance, no particles are detectable in the column effluent. As clearly visible in the linear representation, the appearance of suspended colloids coincides with the decrease in ionic strength. The particle concentration rises sharply to its maximum value, and thereafter the particle concentration decreases gradually. The double-logarithmic representation emphasizes two primary features. First, while no particles are detectable up to one pore volume, a steep raise in the colloid concentration is observed thereafter, peaking at about 2-3 pore volumes. This observation is in agreement with the theoretical predictions for the case of a non-retarded disturbance within a dispersive medium (middle panel of Fig. 1.2).

The second key feature pertains to the long-term behaviour. Particle release can be observed over nearly the entire experimental window of 1000 pore volumes, which corresponds to an experimental duration of about 2 months. The colloid concentration decays very gradually over several orders of magnitude in time, and finally approaches the detection limit (dotted line in Fig. 1.3). The linear appearance on the doubly logarithmic plot suggests a power-law decay. This observation is in sharp contrast with the expected exponential pattern based on a colloid population characterized with a single release rate coefficient. For illustration, the best fit assuming such a model is included in Fig. 1.3, *right* (solid line). Obviously, such a model fails to represent the experimental data.

The non-exponential release behavior can be understood, once one realizes the heterogeneity of a natural system. The wide variability concerning morphology and chemical composition of the particles, but also the topological and structural differences of the attachment environments can be represented in terms of a broad distribution of particle release rate coefficients. This heterogeneity of the colloidal particles initially present in the system results in a characteristic release pattern. Similarly to the experiment shown in Fig. 1.3, a power-law type dependence of the suspended particle concentration as a function of elution time was observed over a very wide range of solution conditions and for all systems investigated (Grolimund and Borkovec 1999; Grolimund et al. 2001b; Grolimund and Borkovec 2001; Grolimund and Borkovec 2006). Such release patterns suggest a considerable long-term potential for the release of colloidal particles.

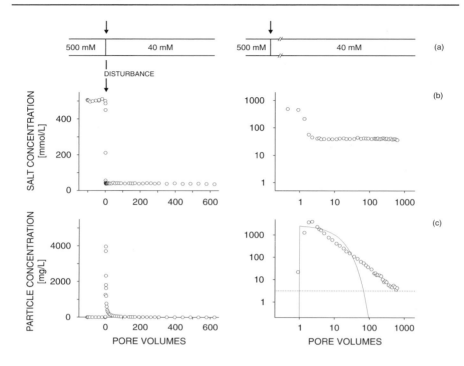

Fig. 1.3 Mobilisation of colloidal particles from a natural porous medium under controlled pH 5.25 (malonate buffer). Release is induced by a chemical disturbance (reduction in ionic strength). a) Feed solutions. b) Effluent concentration of sodium. c) Concentration of colloidal particles suspended in the outflow. The solid line corresponds to a calculation assuming a single uniform colloid population. The dotted line indicates the detection limit.

Two examples demonstrating the importance of chemical conditions on the mobilisation of colloids are depicted in Fig. 1.4. The left panel shows the influence of varying the salt concentration. Colloid mobilisation is strongly enhanced in case of reduced ionic strength. The observed average release rates also suggest an inverse power-law dependence concerning ionic strength. For different soils or aquifer materials, an exponent in the order of 2 to 3 has been observed (Grolimund and Borkovec 1999; Grolimund et al. 2001b; Grolimund and Borkovec 2006). As a consequence, colloid release is sensitively coupled to the ionic strength. As an illustration, consider the case depicted in Fig. 1.4 *left*. A reduction in the ionic strength by a factor of four (i.e., from 80 mM down to 20 mM) enhances the particle release by a factor of 20. Consequently, fluctuations in the pore water chemistry will have a profound effect on the concentration of suspended particles and a strong coupling between the reactive transport pattern and colloid release pattern must be expected. While the amount of available colloids is also important, the release intensity is mainly triggered by the chemical conditions established following a chromatographic front.

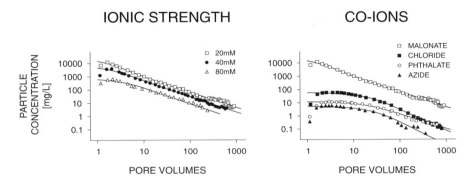

Fig. 1.4. Effects of solution chemistry on colloid release in a sodium dominated porous medium. *left*) Impact of ionic strength under controlled pH 5.25 (malonate buffer). Particle release was initiated by lowering the ionic strength to 20, 40, and 80 mM. *right*) Effect of different co-ions. In the presence of different co-ions, particle release was initiated by lowering the ionic strength to 20 mM. Solid lines correspond to calculations based on power law distributions of release rate coefficients. For details see Grolimund and Borkovec (2001).

One must further realize that anionic ligands have a major effect on the release process. This finding is illustrated in Fig. 1.4 *right*. For otherwise identical systems but different in the type of ligand present, release rates deviate by four (!) orders of magnitude. Unfortunately, however, other experimental investigations studying the impact of small organic and inorganic ligands on dispersion phenomena are virtually nonexistent. Inorganic anions and small organic acids are known to strongly interact with natural oxide and mineral surfaces and to adsorb at the solid-liquid interface. In the context of colloidal stability and dispersion, the interaction of small organic acids with surfaces is known to affect the charging behavior of the corresponding solid phases considerably. Thus it must be expected that any changes in the surface chemistry will also have a profound effect on the release pattern.

Several anionic ligands are currently under consideration or even already under usage within various *in-situ* remediation strategies – most prominent example being EDTA. Such complexing agents bare a considerable potential of enhanced particle mobilisation. Consequently, possible drawbacks of the usage of EDTA for 'soil-washing' include all potential hazards due to particle mobilisation, including structural damage and clogging of the porous medium (see below). These effects could reduce the efficiency of remediation treatment based on washing technologies substantially.

So far, systems dominated by sodium have been discussed. As a next step in progressing towards understanding complex natural conditions, systems dominated by divalent cations will be investigated. Fig. 1.5 compares the release pattern observed for the three cases of (i) a solution of a monovalent cation (Na^+), (ii) solution of a divalent cation (Ca^{2+}), and (iii) a mixture of monovalent and divalent cations (Na^+/Ca^{2+}).

The situation, when the water chemistry is dominated by sodium as a monovalent cation, has been addressed in the previous section in detail. The case of a porous medium saturated with a divalent cation deviates substantially from the sodium dominated system (compare the left and middle panel in Fig. 1.5). Despite a most extreme reduction in ionic strength, no colloids could be detected at the column outlet in the system dominated by the divalent cations. The presence of Ca^{2+} suppresses the release of colloids completely. This phenomenon is probably related to ion-ion correlation forces induced by multivalent ions (Kjellander et al. 1988; Quirk 1994). Although there is little mechanistic understanding of the process, experimental observations point towards a strong impact of the valency of cations on the release behavior.

In the case of the simultaneous presence of cations with different valencies, a complex behavior results. The colloid mobilisation during a classical cation exchange experiment is illustrated in Fig 1.5, *right*. Initially, the porous medium is saturated with monovalent cations. Accordingly, a reduction in ionic strength leads to the observation of a normality front and results in the coupled onset of colloid release. However, particle release ends abruptly with the arrival of the retarded exchange front (compare course of the calcium concentration, Fig. 1.5b, *right*). This phenomenon is in contrast to the extended tailing observed in the sodium-dominated system, but it can be related to the inhibition of particle release by the presence of Ca^{2+} in solution. Such an argumentation, however, is challenged by the second part of the experiment. Following a second disturbance, changing the feed solution back to a sodium solution, release of particles is observed despite the presence of calcium in solution.

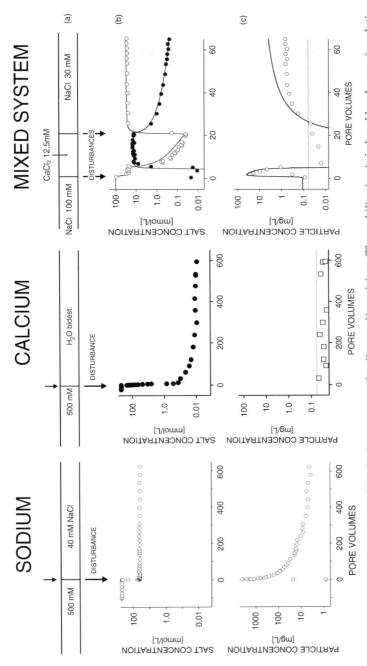

Fig. 1.5. Release of colloidal particles in the presence of sodium and/or calcium. The mobilization is induced by changes in solution chemistry (normality and/or cation composition). a) Input sequences of feed solutions. b) Effluent concentrations of major cations. c) Colloid concentration measured at the column outlet. The case of a sodium saturated system (*left*), a divalent system (*middle*), and the case of the simultaneous presence of sodium and calcium (*right*) are compared. Solid lines in the right panel correspond to model calculations (Grolimund and Borkovec 2006).

A gradual increase in colloid concentration along with a smooth decrease in the Ca^{2+} concentration can be established. Two important observations can be made. First, following the disturbance particle release was initiated although the ionic strength is increased (25mM to 30mM, Fig. 1.5a). Secondly, particles are released despite a calcium concentration which is almost two orders of magnitude higher compared to the calcium-saturated system where no release was observed (Fig. 1.5, *middle*). Considering the conditions at the colloid-solution interface yields a possible explanation of these two observations.

In the calcium-dominated system, the colloid surface is almost exclusively saturated by Ca^{2+} ions. In the mixed Na^+/Ca^{2+} system, however, monovalent sodium ions and Ca^{2+} ions compete for the surface sites. At the later stage of the cation exchange experiment shown in Fig. 1.5, *right*, the fraction of surface sites occupied by calcium decreases gradually to a value of approximately 0.3. Accordingly, the conditions at the colloid-solution interface slowly approach the conditions of a sodium-dominated system. One may conclude that the intensity of particle mobilisation is also controlled by the fractional concentration of divalent cations at the colloid-solution interface.

The examples shown in Fig. 1.4 and 1.5 clearly document the sensitivity of the release process towards the prevalent chemical conditions. This specific coupling between mobilisation of colloidal particles and multi-component transport phenomena of cations and co-ions governs the concentration pattern of suspended particles, not only in model systems but also in the subsurface.

1.4 Transport and Persistence

After mobilisation, the lifetime of suspended colloidal particles in porous media is determined by convective transport, particle (re-) deposition, and the aggregation behavior.

The morphology and structure of the porous medium influences the convective transport of the mobile particles, which can be distinctly different from a conservative tracer. An example is shown in Fig. 1.6. Two differences between the colloidal particles and the conservative tracer are readily apparent: (i) colloidal particles in the soil column travel considerably faster than a non-reactive tracer (earlier breakthrough), and (ii) only a fraction of the injected particles is recovered at the column effluent (reduced integrated area). The phenomenon of faster transport of colloidal particles is known as size exclusion within classical chromatography and is based on the finite size of the particles. In contrast to a conservative tracer, a considerable fraction of the pore space is not accessible for the particles due to geometrical restrictions.

The difference in size of tracer versus colloids is also reflected in the observed discrepancy regarding tracer and colloid dispersivities. The dispersivity as a function of pore water velocity is depicted in Fig. 1.6, *right*. In case of the colloids, the observed dependency indicates pure mechanical dispersion. For the tracer, the observed power-law behavior points towards a combination of holdup dispersion and advective dominated dispersion. Most of the inter-aggregate pore space is not accessible to the colloids as pore diameters are smaller than the average particle size. For the tracer molecules, the solid phase must be viewed as a porous medium with two very different pore sizes (i.e., double porosity) (Grolimund et al. 1998).

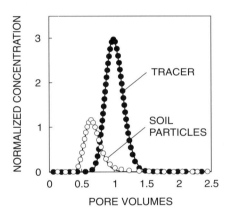
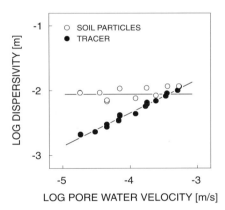

Fig. 1.6. General particle transport behavior in a packed soil column resulting from pulse input experiments. Breakthrough curves of the natural soil colloidal particles compared to an inert tracer. Solid lines correspond to model calculations (Grolimund et al. 1998) (*left*). Influence of flow velocity on tracer and colloid dispersivity. Solid line represents eye-guides *(right)*.

The second distinct feature of the particle breakthrough curves in Fig. 1.6, the reduced peak height and area of the normalized particle breakthrough, can be attributed to immobilisation of colloidal particles on the surfaces of soil grains due to particle deposition (physico-chemical filtration). The process of deposition corresponds to a temporary sink for colloidal particles. Contrary to particle mobilisation, the deposition process initially follows first order kinetics (Grolimund et al. 1998). Particle (re-)

deposition turned out to be dominantly influenced by solution chemistry as well as the surface chemical properties of the colloidal particles and of the porous media. As an example, the influence of the electrolyte concentration, and counterion valence on the deposition rate coefficient for *in-situ* mobilized soil particles in their parent porous medium is shown in Fig. 1.7a. One observes that the ionic strength in general and the valence of the present counter ions have a profound influence on the resulting deposition rates. These and other effects of chemical and physical system parameters on the deposition behavior can be explained – at least qualitatively – with existing concepts and theories (Elimelech et al. 1995). For example, filtration theory, although conceptualized based on ideal model systems, proved capable to predict quantitatively the transport behavior of *in-situ* mobilized colloidal particles within their parent, natural porous media (Grolimund et al. 1998).

Since the nature of the anions had an important effect on particle mobilisation, their influence was also investigated for colloid deposition (Fig. 1.7, *left*). While the release process is substantially different for chloride compared to azide (Fig. 1.7, *right*), particle deposition measured in the same system was identical for both ligands. We suspect therefore that the nature of anions has much weaker influence on particle deposition than on particle release.

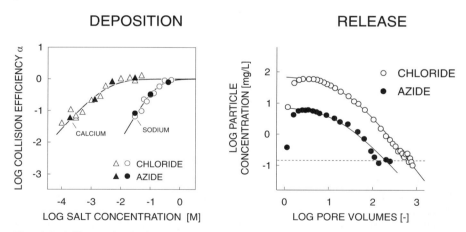

Fig. 1.7. Effects of solution chemistry on colloid deposition and release. Effect of salt concentration, counterion valence (Na$^+$ vs. Ca^{2+}), and type of co-ion on deposition rate coefficients (*left*). Effect of different co-ions on colloid release (*right*). In contrast to colloid deposition, which was not affected by the presence of azide, particle release is strongly suppressed.

Clearly, the molecular-level structure of the colloid-solution interface is the key to elucidate this difference between deposition and release. Particle release rates are directly linked to the depth of the primary minimum. Consequently, the release behavior is dominated by short-range forces and will be affected specifically by the nature of the adsorbed molecules. In case of particle deposition, the height of the potential barrier is rate limiting. While particle deposition does also depend on the surface composition, this process is mainly influenced by forces at larger separations, which are rather of electrostatic nature and less susceptible to ion specific effects.

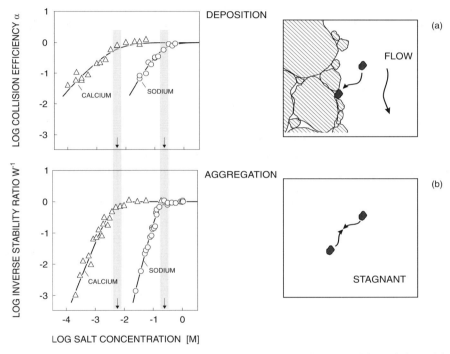

Fig. 1.8. Comparison of the deposition kinetics of mobile colloidal particles with their aggregation kinetics. Relative rate constants are plotted as a function of salt concentration for NaCl and CaCl$_2$ in a doubly logarithmic fashion. (a) Collision (attachment) efficiency for deposition, and (b) inverse stability ratio for aggregation. Solid lines are best fits. For further details see Grolimund et al. (2001a).

The aggregation to larger conglomerates and subsequent straining (physical filtration) represents another process limiting the persistence of mobile particles in subsurface porous media. Dynamic light scattering techniques were used to study the aggregation behavior of *in-situ* mobilized colloidal particles over a wide range of solution conditions. The results of such aggregation kinetics measurements are depicted in Fig. 1.8 and they can easily be compared to the previously discussed particle deposition kinetics. The analogy between deposition and aggregation kinetics is evident. However, the similarity between aggregation and deposition is not too surprising, since the colloidal particles were obtained through release from the parent soil matrix, which also serves as the natural porous medium in deposition. Therefore, the interaction between suspended colloidal particles can be expected to be similar to the interaction of particles with the stationary solid matrix. Consequently, particle deposition in mono-ionic and mixed cation systems can be effectively studied via colloidal stability measurements by dynamic light scattering (Grolimund et al. 2001a).

1.5 Structural Changes

The mobilisation and subsequent redeposition or filtration of colloidal particles can have secondary effects such as changed hydraulic properties, or changed chemical reactivity of the porous media due to the translocation of reactive surface area. Such phenomena are of particular importance in the context of 'pump and treat' remediation techniques. Flushing with chemically active agents ('soil washing') corresponds to one of the most widely applied in-situ remediation strategies for contaminated subsurface systems. As was discussed earlier, complexing agents may enhance the susceptibility for particle mobilisation dramatically (Fig. 1.4, *right*). Nevertheless, the potential problem of structural damage is commonly not addressed within the performance assessment of remediation strategies. However, evidence for the modification of hydraulic properties and structural damage due to mobilized colloids can be found within various scientific fields. For example, mobile colloidal particles have been recognized to play an important role in the structural damage of sodic soils (Quirk and Schofield 1955; Bresler et al. 1982; Birkeland 1984; Quirk 1994), and are suspected to be responsible for changed hydraulic properties of subsurface porous media after artificial recharge (Shainberg et al. 1980; Goldenberg et al. 1984; Wiesner et al. 1996). Furthermore, formation plugging by mobilized colloidal particles represent a major problem in secondary oil recovery and has been subject of many investigations (Jones 1964; Mungan 1965; Reed 1972; Muecke 1979; Shainberg et al. 1980; Goldenberg et al. 1984; Khilar and Fogler 1987).

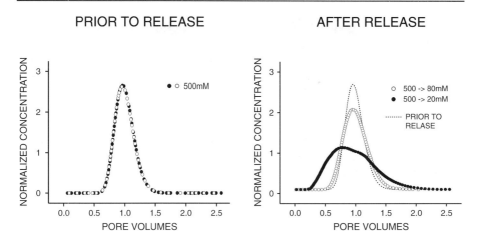

Fig. 1.9. Structural damage. Comparison of tracer pulse breakthrough prior to (*left*) and subsequent to colloid mobilisation (*right*). Following mobilisation, the distorted breakthrough pattern is indicative of increased dispersivities and non-idealities such as the formation of dead-end pores and the clogging of pore space.

Comparative tracer studies before and after mobilisation events provide additional experimental evidence for structural damage following colloid release. The outcome of such a tracer study is shown in Fig. 1.9. The left panel depicts pulse tracer breakthrough curves recorded prior to the mobilisation of particles. Two independent column experiments yield identical tracer breakthrough pattern. The time-dependent tracer concentration measured at the column outlet is in close agreement with calculations based on the classical convection-dispersion equation. The corresponding tracer transport patterns for the very same porous medium – but after colloid mobilisation – are shown in the right panel of Fig. 1.9. As immediately apparent, the tracer breakthrough curves are considerably distorted. A more pronounced degree of distortion is observed for increased release intensities. The shape of the distorted breakthrough pattern is indicative of the formation of dead-end pores and the clogging of pore space.

1.6 Colloid Facilitated Transport

Numerous field-scale and laboratory-based studies have documented the importance of colloid facilitated transport (Champlin and Eichholz 1968; Champ et al. 1982; Vinten and Nye 1985; Buddemeier and Hunt 1988; Fauré et al. 1996; Saiers and Hornberger 1996; Roy and Dzombak 1997;

de Jonge et al. 1998; Kersting et al. 1999). This transport mechanism was also observed to be important in the same natural porous medium, which was used to investigate the fundamental particle release and deposition phenomena as described in the previous sections (Grolimund et al. 1996; Grolimund and Borkovec 2005). An extended transport model considering colloidal phenomena was formulated by compiling the knowledge gathered on specific processes and phenomenon into an integrated framework. Clearly, the three processes of particle mobilisation, deposition and aggregation occur simultaneously. However, the relative importance of each of these processes is strongly dependent on the present chemical and physical conditions. Any extended reactive transport model attempting to include colloidal particles has to consider the strong coupling between colloidal phenomena and reactive transport. In the following, only a short summary of the conceptual framework will be given. Further details can be found elsewhere (Grolimund and Borkovec 2005; Grolimund and Borkovec 2006).

While developing such an extended reactive transport model, an appropriate formulation of the *in-situ* release phenomena represents the key obstacle. The pronounced non-exponential release behavior, the finite supply of colloidal particles, and – most challenging – the strong dependence of the observed release kinetic on chemical system parameters have to be considered. In the present model, the finite supply and the non-exponential character of the particle release are captured by assuming a discrete distribution of colloid populations with the density distribution of release rate coefficients following a power distribution (Grolimund and Borkovec 1999; Grolimund et al. 2001b). Concerning the dependence on chemical conditions, two key parameters are considered: (i) the ionic strength of the pore water and (ii) the fraction of surface sites occupied by divalent cations. In both cases power-law relations are used to represent the dependence on these parameters (Grolimund and Borkovec 2005; Grolimund and Borkovec 2006).

Particle transport is described with the classical filtration theory, including enhanced particle velocity, modified dispersion coefficients of the released particles, and deposition triggered by solution chemistry. However, the model does not consider changes in the physical structure of the porous medium due to deposition, filtration, or aggregate breakdown.

Chemical reactions taking place in solution and at the solid-solution interfaces are formulated in terms of kinetic rate laws. The required forward and backward reaction rates are obtained based on the analysis of breakthrough curves. Concerning interfacial reactivity, the loss in total surface area due to colloid mobilisation is considered explicitly.

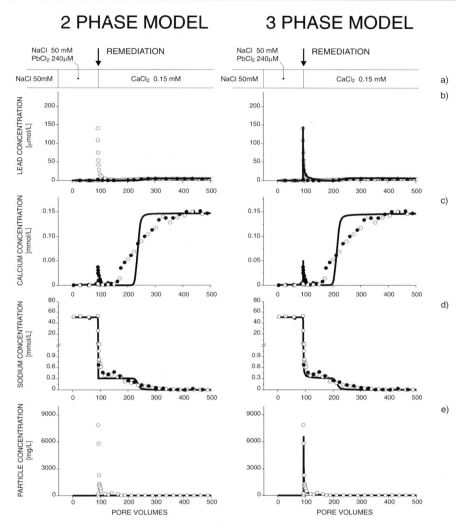

Fig. 1.10. Contamination-remediation scenario. Column experiments demonstrating facilitated transport of Pb^{2+} by *in-situ* mobilized colloids. Colloid facilitated transport dominates the transport of Pb^{2+} for at least 50 pore volumes after the mimicked remediation action. Solid lines are calculations based on a classical 2-phase and extended 3-phase reactive transport model, respectively (Grolimund and Borkovec 2005; Grolimund and Borkovec 2006). Without the consideration of colloid facilitated transport, the observed reactive transport pattern can not be reproduced.

A transport model based on these ideas was developed, and it has proven to be successful in the description of experiments with coupled multicomponent transport and particle mobilisation in a natural porous medium (Grolimund and Borkovec 2005; Grolimund and Borkovec 2006). One example is given in Fig. 1.10. The sequence of feed solutions (Fig. 1.10a) mimics a typical contamination-remediation scenario. A contaminated salty plume is infiltrating an aquifer system. Thereby, different chromatographic fronts develop. More mobile constituents of the plume arrive at the sampling well, while highly retarded chemicals (such as the model contaminant Pb^{2+}) migrate only part of the distance between dump site and sampling well. After the remediation of the source of contamination, the aquifer is again solely infiltrated by pristine groundwater. Contrary to expectations, immediately after the remediation, a pulse of contaminants is observed at the sampling well (at ~95 pore volumes). At a later stage of the experiment, a second, but much broader, contaminant pulse but of lower concentration is observed. Concurrently with the first pollution peak, suspended particles are detected. The differentiation between truly dissolved contaminants and colloidal contaminant species (open and closed circles in Fig. 1.10b) proves the association of contaminants with the suspended colloids. Clearly, a classical 2-phase reactive model is not able to reproduce the observed reactive transport pattern (Fig. 1.10, *left*). In contrast, the consideration of carrier colloids within a 3-phase model yields satisfactory agreement between model calculations and experimental results (Fig. 1.10, *right*).

The presented 3-phase model was applied to an extensive set of additional mobilisation experiments and contaminant migrations studies (Grolimund and Borkovec 2005; Grolimund and Borkovec 2006). This modelling exercise clearly demonstrated the pronounced impact of the solution as well as surface composition on colloidal phenomena. The model is able to predict the observed breakthrough pattern of complex contaminant transport experiments, including the phenomena of enhanced contaminant transport by *in-situ* mobilized colloidal particles.

1.7 Conclusions and Outlook

In this chapter, the fundamental phenomena relevant to colloid facilitated transport have been discussed. The focus was on *in-situ* mobilisation (*"generation"*) of colloidal particles, their transport, deposition, and aggregation (*"persistence"*). Furthermore, we have highlighted the important effects of solution and interfacial chemistry (*'contaminant–colloid interaction'*).

Let us stress the two most important observations here. Firstly, the generation and persistence of colloidal particles are highly sensitive to chemical conditions. As a consequence, suspended colloid concentrations are coupled to the non-linear chromatographic transport behavior of chemical species present in natural porous media. The chemical susceptibility of particle mobilisation turned out to be distinctly different with respect to particle deposition. Particle release shows much more pronounced dependence on the chemical conditions at the solid-liquid interface that particle deposition. This difference is clearly reflected in the marked effect of counter-ions on particle release.

Secondly, physical and chemical heterogeneities influence the nature of the particle release and deposition process strongly. The non-exponential character of the release process represents the most clear example of this behavior. Colloidal particles in a natural porous media cannot be represented by a single, homogeneous population, but a distribution of particle populations with different properties must be considered.

The attempt to model reactive transport experiments involving the release of colloidal particles clearly discloses the necessity of considering the chemical susceptibility as well as different types of heterogeneities explicitly.

The present combined approach of experimental investigations and mathematical modelling corresponds to a powerful strategy to achieve a detailed understanding of the chemical or physical system parameters, which control particle mobilisation, transport, and persistence under field conditions. Such combined efforts result in an improved ability to understand and predict the susceptibility of natural systems for hazardous phenomena induced by mobilized colloidal particles such as enhanced transport of contaminant associated with mobile particles.

Over the past years, remarkable progress in the understanding of fundamental colloidal phenomena of importance in the context of reactive transport has been achieved (Ryan and Elimelech 1996; Kretzschmar et al. 1999; Sen and Khilar 2006) . Particle deposition has been studied by many researchers, and a relatively good level of understanding has been achieved. However, particle release has been studied in much lesser detail, and the mechanistic details of the process remain obscure. To conclude this chapter, a selection of key challenges and research needs is provided in the following:

- An improved understanding of particle release necessitates detailed information about the solid-solution interface down to the molecular level. An intensification of research efforts directed towards an improved molecular level understanding is required.

- The insufficient knowledge about interaction forces acting between the larger grains and the attached colloids at small separation distances represents an important issue requiring further clarification.

- Experimental observations point towards a strong impact of the valency of cations on the release behavior. The role of ion-ion correlation forces should be elucidated in more detail.

In addition to the aforementioned mechanistic studies, empirical but systematic investigations can equally contribute to an improved understanding of the relevant processes.

- Studies investigating various release and deposition phenomena in the same experimental system and under identical conditions would be beneficial.

- While the valency of the dominant counter-ion was shown to play a dominant role in particle release, ion specific effects of different counter-ions of the same valency have not yet been investigated.

- Experimental investigations focusing on the impact of co-ions (i.e., small organic and inorganic ligands) on dispersion phenomena are virtually not existent.

- Structural damage of porous media due to particles mobilisation represents a suite of multiple colloidal phenomena of high relevance.

The chemical and physical complexity of natural systems requests a profound understanding of the impact of multicomponent effects. To understand the susceptibility of colloidal phenomena regarding chemical fluctuations can be regarded as one of the key scientific challenges in the context of colloid facilitated transport.

References

Appelo CAJ, Postma D (1996) Geochemistry, groundwater and pollution. A.A. Balkema, Rotterdam

Birkeland PW (1984) Soils and geomorphology. Oxford University Press, New York

Bresler E, McNeal BL, Carter DL (1982) Saline and sodic soils. Principles-dynamics-modeling. Springer-Verlag, Berlin

Buddemeier RW, Hunt JR (1988) Transport of colloidal contaminants in groundwater: Radionuclide migration at the Nevada Test Site. Appl Geochem 3:535-548

Champ DR, Merritt WF, Young JL (1982) Potential for the rapid transport of plutonium in groundwater as demonstrated by core column studies. In: W Lutze (ed), Scientific basis for radioactive waste management. Elsevier, pp 745-754

Champlin JBF, Eichholz GG (1968) The movement of radioactive sodium and ruthenium through a simulated aquifer. Water Resour Res 4:147-158

Corapcioglu MY, Jiang S (1993) Colloid-facilitated groundwater contaminant transport. Water Resour Res 29:2215-2226

Dagan G (1989) Flow and transport in porous formations. Springer, New York

de Jonge H, Jacobsen OH, de Jonge LW, Moldrup P (1998) Particle-facilitated transport of prochloraz in undisturbed sandy loam soil columns. J Environ Qual 27 (6):1495-1503

Elimelech M, Gregory J, Jia X, Williams RA (1995) Particle deposition and aggregation. Measurement, modelling, and simulation. Butterworth-Heinemann, Oxford

Fauré MH, Sardin M, Vitorge P (1996) Transport of clay particles and radioelements in a salinity gradient: experiments and simulations. J Contam Hydrol 21:255-267

Goldenberg LC, Magaritz M, Amiel AJ, Mandel S (1984) Changes in hydraulic conductivity of laboratory sand-clay mixtures caused by a seawater-freshwater interface. J Hydrol 70:329-336

Grolimund D, Borkovec M (1999) Long term release kinetics of colloidal particles from natural porous media. Environ Sci Technol 33 (22):4054-4060

Grolimund D, Borkovec M (2001) Release and transport of colloidal particles in natural porous media. 1. Modeling. Water Resour Res 37 (3):559-570

Grolimund D, Borkovec M (2005) Colloid-facilitated transport of strongly sorbing contaminants in natural porous media: mathematical modeling and laboratory column experiments. Environ Sci Technol 39 (17):6378-6386

Grolimund D, Borkovec M (2006) Release of colloidal particles in natural porous media by monovalent and divalent cations. J Contam Hydrol 87 (3-4):155-175

Grolimund D, Elimelech M, Borkovec M (2001a) Aggregation and deposition kinetics of mobile colloidal particles in natural porous media. Colloids and Surfaces, A: Physicochemical and Engineering Aspects 19 (1-2):179-188

Grolimund D, Barmettler K, Borkovec M (2001b) Release and transport of colloidal particles in natural porous media. 2. Experimental results and effects of ligands. Water Resour Res 37 (3):571-582

Grolimund D, Borkovec M, Barmettler K, Sticher H (1996) Colloid-facilitated transport of strongly sorbing contaminants in natural porous media: a laboratory column study. Environ Sci Technol 30 (10):3118-3123

Grolimund D, Elimelech M, Borkovec M, Barmettler K, Kretzschmar R, Sticher H (1998) Transport of in situ mobilized colloidal particles in packed soil columns. Environ Sci Technol 32 (22):3562-3569

Gschwend PM, Reynolds MD (1987) Monodisperse ferrous phosphaste colloids in an anoxic groundwater plume. J Contam Hydrol 1:309-327

Jones FO (1964) Influence of chemical composition of water on clay blocking of permeability. J Petrol Technol 16:441-446

Jury WA, Roth K (1990) Transfer functions and solute movement trough soils: Theory and applications. Birkhäuser, Basel

Kersting AB, Efurd DW, Finnegant DL, Rokop DJ, Smith DK, Thompson JL (1999) Migration of plutonium in ground water at the Nevada Test Site. Nature 397:56-59

Khilar KC, Fogler HS (1987) Colloidally induced fines migration in porous media. Rev Chem Eng 4 (1&2):41-108

Kjellander R, Marcelja S, Pashley RM, Quirk JP (1988) Double-layer ion correlation forces restrict calcium clay swelling. J Phys Chem 92:6489-6492

Knox RC, Sabatini DA, Canter LW (1993) Subsurface transport and fate processes. Lewis Publishers, Boca Raton

Kretzschmar R, Borkovec M, Grolimund D, Elimelech M (1999) Mobile subsurface colloids and their role in contaminant transport. Adv Agron 66:121-193

Lenhart JJ, Saiers JE (2003) Colloid mobilization in water-saturated porous media under transient chemical conditions. Environ Sci Technol 37(12):2780-2787

Lichtner PC, Steefel CI, Oelkers EH (eds) (1996) Reactive transport in porous media. Reviews in mineralogy, 34. The Mineralogical Society of America, Washington, DC

Muecke TW (1979) Formation fines and factors controlling their movement in porous media. J Petrol Technol 31 (2):144-150

Mungan N (1965) Permeability reduction through changes in pH and salinity. J Petrol Technol 17:1449 -1453

Nightingale HI, Bianchi WC (1977) Ground-water turbidity resulting from artificial recharge. Ground Water 15:146-152

Quirk JP (1994) Interparticle forces: A basis for the interpretation of soil physical behavior. Adv Agron 53:121-183

Quirk JP, Schofield RK (1955) The effect of electrolyte concentration on soil permeability. J Soil Sci 62:163 -178

Reed MG (1972) Stabilization of formation clays with hydroxy-aluminium solutions. J Petrol Technol :860-864

Roy SB, Dzombak DA (1997) Chemical factors influencing colloid-facilitated transport of contaminants in porous media. Environ Sci Technol 31 (3):656-664

Roy SB, Dzombak DA (1998) Sorption nonequilibrium effects on colloid-enhanced transport of hydrophobic organic compounds in porous media. J Contam Hydrol 30:179-200

Ryan JN, Elimelech M (1996) Colloid mobilization and transport in groundwater. Colloids and Surfaces, A: Physicochemical and Engineering Aspects 107:1-56

Saiers JE, Hornberger GM (1996) The role of colloidal kaolinite in the transport of cesium through laboratory sand columns. Water Resour Res 32 (1):33-41

Sardin M, Schweich D, Leij FJ, van Genuchten MT (1991) Modeling the nonequilibrium transport of linearly interacting solutes in porous media: a review. Water Resour Res 27 (9):2287-2307

Sen TK, Khilar KC (2006) Review on subsurface colloids and colloid-associated contaminant transport in saturated porous media. Adv Coll Interf Sci 119:71-96

Sen TK, Shanbhag S, Khilar KC (2004) Subsurface colloids in groundwater contamination: A mathematical model. Colloids and Surfaces, A: Physicochemical and Engineering Aspects 232 (1):29-38

Shainberg I, Rhoades JD, Prather RJ (1980) Effect of low electrolyte concentration on clay dispersion and hydraulic conductivity of a sodic soil. Soil Sci Soc Am J 45:273-277

van de Weerd H, Leijnse A, van Riemsdijk WH (1998) Transport of reactive colloids and contaminants in groundwater: effect of nonlinear kinetic interactions. J Contam Hydrol 32:313-331

van der Lee J, Ledoux E, de Marsily G, de Cayeux MD, van de Weerd H, Fraters B, Dodds J, Rodier E, Sardin M, Hernandez A (1994) A bibliographical review of colloid transport through the geosphere, European Commission, Luxembourg.

Vinten AJA, Nye PH (1985) Transport and deposition of dilute colloidal suspensions in soils. J Soil Sci 36:531-541

Wiesner MR, Grant MC, Hutchins SR (1996) Reduced permeability in groundwater remediation systems: role of mobilized colloids and injected chemicals. Environ Sci Technol 30:3184-3191

Wu Q, Borkovec M, Sticher H (1993) On particle-size distribution in soils. Soil Sci Soc Am J 57:883-890

2 Influence of Na-bentonite Colloids on the Transport of Heavy Metals in Porous Media

George Metreveli, Fritz H. Frimmel

Engler-Bunte-Institut, Chair of Water Chemistry, Universität Karlsruhe (TH), Engler-Bunte-Ring 1, D-76131 Karlsruhe, Germany
e-mail: george.metreveli@ebi-wasser.uka.de

2.1 Abstract

In this work, the influence of Na-bentonite colloids on the transport of Cu, Pb and Zn in porous media was investigated. For the transport experiments a "short pulse" laboratory column system was used. Quartz sand served as column packing material. The metal solutions were injected into the column in the presence and absence of colloids. The quantification of metals at the column outlet was carried out by coupling the column system with an inductively coupled plasma mass spectrometer (ICP-MS). The determination of Na-bentonite colloids was done online by means of a UV detector and the ICP-MS system. Characterisation of the colloid-metal interactions was based on sorption experiments and modelling calculations carried out at pH values of 5 and 7. Furthermore the stability of Na-bentonite colloids was determined in titration experiments.

The titration experiments showed that in the range $4 \leq pH \leq 12$ Na-bentonite colloids remained stable. At lower pH values ($2 < pH < 4$), bentonite particles aggregated rapidly. In the sorption experiments a small pH-dependence of the copper and lead adsorption onto bentonite could be detected. For zinc no significant influence of pH value on the sorption behaviour was found. The results of modelling calculations showed a very high amount of adsorbed heavy metals and were therewith in good agreement with the experimental data. In the transport experiments, Na-bentonite colloids showed high mobility at pH value of 7. At pH 5, the deposition of Na-bentonite particles was observed, which was different in the presence of different heavy metals and increased in the following order: Pb < Zn < Cu. In the absence of Na-bentonite colloids no significant transport of heavy metals was found. In the presence of Na-bentonite a colloid facilitated transport for all heavy metals was observed. The amount of transported heavy metals increased in the following order: Pb < Cu < Zn. The colloid facilitated transport of heavy metals was influenced by the particle deposition and competitive sorption processes between Na-bentonite and column packing material.

2.2 Introduction

Several investigations and experimental studies show that colloidal particles can facilitate the transport of different pollutants in soils (McCarthy and Zachara 1989, Ryan and Elimelech 1996, Kretzschmar et al. 1999). After infiltration of rainwater in urban areas, the pollutants, adsorbed or bound onto the mobile colloids, can discharge into the groundwater. As a consequence, a great risk for drinking water may result, because the groundwater is often used as raw water for drinking water. The possible emission sources of the pollutants are, for example road traffic, building roofs, contaminated sites or waste disposal areas.

The most occurring subsurface colloids are inorganic nano particles (clay minerals, metal oxides or oxyhydroxides, silica, carbonates), natural organic matter (NOM), bacteria and viruses (McCarthy and Zachara 1989, Kretzschmar et al. 1999). The colloidal particles usually have a large specific surface and are in natural aquatic systems particularly mobile. They can adsorb or bind the different contaminants and by this transport cationic metal species, like heavy metals (Grolimund et al. 1996, Roy and Dzombak 1997, Kretzschmar et al. 1999) as well as radio nuclides (Kretzschmar et al. 1999, Saiers and Hornberger 1999, Zhuang et al. 2003) and organic micro pollutants of a broad variety (Vinten et al. 1983, Roy and Dzombak 1997, Kretzschmar et al. 1999).

The colloid facilitated transport of pollutants in soils is depending on the following processes: adsorption or binding of contaminants onto the mobile colloids, interactions between colloids and soil matrix (colloid deposition, blocking and ripening effects, colloid release) and formation of colloid aggregates. These processes are influenced by different factors, like interfacial processes, solution chemistry, hydraulic conditions and presence of various aquatic substances. The chemical conditions play an important role for the association of pollutants with the mobile colloids and the transport trough porous media. Sorption of heavy metals onto the colloidal particles increases generally with increasing pH value (Christl and Kretzschmar 2001). At low salt concentrations, because of the dominated repulsion between colloids and the immobile phase the colloids are highly mobile. Laboratory scale column experiments showed that kaolinite colloids accelerate transport of cesium at low ionic strength (Saiers and Hornberger 1999). Increasing cation concentration in liquid phase favours the increase of attracting electrostatic forces between colloids and grains of the solid matrix, and as a consequence leads to an increase of the deposition rate (Elimelech and O`Melia 1990, Kretzschmar and Sticher 1997, Kretzschmar et al. 1997, Grolimund et al. 1998, Kretzschmar et al. 1999). The colloid deposition depends also on the charge of the cations. In case, the system is dominated by divalent cations (Ca^{2+}) the colloid deposition rate is much higher then for systems with monovalent cations (Na^+) (Grolimund et al. 1998, Kretzschmar et al. 1999). Due to the coagulation, the mobility of humic acid decreases with the increasing Ca^{2+} concentration (Weng et al. 2002). The colloid mobility is also influenced significantly by the pH value of the liquid phase. With increasing solution pH the surfaces of colloids and grains of the solid matrix are more negatively charged. This causes the increase of repulsive forces between colloids and grains and the colloid mobility increases (Litton and Olson 1993, Ryan and Elimelech 1996).

Several studies have shown the influence of NOM on the colloid transport. The adsorption of humic substances onto the oxide surfaces and the edge sites of the clay minerals causes the stabilisation of colloid dispersions and decrease of colloid deposition rate (Kretzschmar et al. 1995, Kretzschmar and Sticher 1997).

Furthermore the change of ion concentration in liquid phase has a great influence on the release of inorganic and organic colloids from solid matrices. The experiments have shown that with decreasing salt concentration (Roy and Dzombak 1996, Grolimund et al. 2001) and increasing pH value (Bunn et al. 2002) the colloid release rate increases. Similar results for the soil organic matter are reported by Oste et al. (2002). The increase of pH value and decrease of Ca^{2+} concentration favours the mobilisation of organic matter from soil material. The release of colloidal particles can result in the mobilisation of associated contaminants (Grolimund et al. 1996, Roy and Dzombak 1997, Flury et al. 2002). Furthermore the blocking and ripening effects play an important role for colloid transport in porous media. If the repulsion forces between colloidal particles (interparticle repulsion) are dominant, the increased amount of deposited particles causes a decrease of the colloid deposition rate. This effect is known in literature as blocking effect (Johnson and Elimelech 1995, Liu et al. 1995, Johnson et al. 1996). In systems with dominating colloid-colloid attraction forces, the multilayer adsorption of colloids onto the surface of solid matrices takes place and the colloid deposition rate increases with increasing amount of deposited particles (ripening effect) (Ryde et al. 1991, Liu et al. 1995, Kulkarni et al. 2005).

In this work the influence of Na-bentonite on the transport of heavy metals (Cu, Pb and Zn) in porous media was investigated. For these investigations a "short pulse" laboratory column system with online metal and colloid detection was used. The transport experiments were carried out under water saturated conditions. The bentonite-metal interactions were characterised by using of sorption experiments and modelling calculations. Furthermore the stability of bentonite colloids was determined.

2.3 Materials and Methods

2.3.1 Substances and Sample Preparation

The sample preparation in all experiments was carried out by using deionised water (Milli-Q Plus, Millipore). The pH value was adjusted with hydrochloric acid (suprapur, Merck) and sodium hydroxide (Merck). Na-bentonite (Ikomont, S&B Industrial Minerals GmbH) was used as inorganic colloid. The Na-bentonite consisted of 85% montmorillonite, 10% other clay minerals, < 2% quartz and < 5% further minerals. The specific surface area of the bentonite particles was 70 m²/g according to manufacturer data. The chemical composition of the Na-bentonite is given in Table 2.1.

Table 2.1. Chemical composition of used materials in % (manufacturer data)

	Na-bentonite	quartz sand
SiO_2	59.9	99.5
Al_2O_3	18.6	0.25
Fe_2O_3	3.4	0.04
MgO	4.5	-
CaO	2.0	-
Na_2O	2.4	-
K_2O	1.1	-
TiO_2	0.4	-
thermal loss	7.2	0.2

Bentonite suspensions were prepared with deionised water at a concentration of 4 g/L and treated in an ultrasonic bath for 15 min. For the removal of large bentonite particles (> 1 μm), the suspensions were centrifuged at 2000 rpm (622 × g) for 15 min (ROTANTA 460 RS, Andreas Hettich GmbH & Co KG). After centrifugation, the supernatant was gained. In the supernatant the particle size distribution was determined by using a Zetasizer Nano ZS (Malvern Instruments, 4 mW He-Ne 633 nm laser) and the total dry residue was quantified according to DIN 38 409-H1-1 Teil 1 (1987). The supernatant was kept prior to use at +8°C in the dark. This supernatant with a bentonite concentration of 1.6 g/L and an average particle size of 280 nm was used as a stock dispersion for the preparation of all bentonite samples.

The quartz sand (F34, Quarzwerke Frechen, mean particle size 200 μm, specific surface area 118 cm²/g) was used as a column packing material in the transport experiments. The chemical composition of the quartz sand is shown in Table 2.1. For removing possibly present organic or inorganic (metal species) contaminants the quartz sand was washed with a solution of sodium hydroxide (c(NaOH) = 0.1 mol/L), a solution of nitric acid (c(HNO$_3$) = 0.1 mol/L) and finally with deionised water. The washed quartz sand was dried at +105°C in a desiccator cabinet (T6060, Heraeus Instruments).

For the preparation of metal stock solutions the following chemicals were used: CuCl$_2$·2H$_2$O (Merck), PbCl$_2$ (Merck), ZnCl$_2$ (Merck). These salts were dissolved in a solution of hydrochloric acid (c(HCl) = 10 mmol/L). The concentration of metals was set to 10 mmol/L. These metal stock solutions were used for the preparation of samples in all experiments.

2.3.2 Titration Experiments

For the stability characterisation of Na-bentonite dispersions the titration experiments were done by means of a Zetasizer Nano ZS and an Autotitrator (MPT-2, Malvern Instruments). The titration was carried out in a plastic container, which was connected through capillary system and peristaltic pump with a folded capillary zeta potential cell (DTS 1060, Malvern Instruments). The bentonite stock dispersion was diluted to a concentration of 200 mg/L with demineralised water. The 10 mL of the diluted bentonite dispersion (200 mg/L) was titrated from pH value of 7 to 2 with 0.1 mol/L and 1 mol/L hydrochloric acid and from pH value of 7 to 12 with 0.1 mol/L sodium hydroxide. The pH value was measured during the titration with a pH electrode. After each pH adjustment, the particle size and electrophoretic mobility was detected three times. The particle size was measured by dynamic light scattering (scattering angle 173°) and the electrophoretic mobility was detected by means of laser doppler electrophorese technique. The zeta potential was calculated from electrophoretic mobility by using of Smoluchowski equation. Between the measurements and during the pH adjustments, the sample was circulated and stirred. Every titration experiment was carried out twice. The titration experiments were controlled fully automatically by software.

2.3.3 Adsorption Experiments and Modelling

Sorption Experiments

Sorption of heavy metals (Cu, Pb and Zn) onto Na-bentonite surface was investigated by detecting of adsorption isotherms. The isotherm samples were prepared from stock dispersion of Na-bentonite and stock solutions of heavy metals in 40 mL glass flasks by dilution with demineralised water. In the isotherm samples, the concentration of Na-bentonite between 1 mg/L and 500 mg/L and the concentration of heavy metals between 10 μmol/L and 50 μmol/L were varied. The samples were equilibrated for 10 days at 25°C and at pH values of 5 and 7. After equilibration, from every sample was three 3.5 mL aliquot samples gained, transferred to the centrifuge tubes and centrifuged by using of ultracentrifuge (Optima TLX, Beckman Coulter) at 35000 rpm (60000 × g) for 3 hours (Accel time from 0 to 5000 rpm 5 min, decel time from 5000 to 0 rpm 2.5 min). After centrifugation, the 1.5 mL supernatant from each tube was gained. The supernatants of same samples from three centrifuge tubes were collected in one tube (4.5 mL), stabilised in 1% nitric acid (suprapur, Merck) and analysed for metal concentration with inductively coupled plasma mass spectrometer (ICP-MS, ELAN 6000, Perkin Elmer).

Sorption Modelling

The sorption of heavy metals (Cu, Pb and Zn) onto Na-bentonite was calculated using the computer software Visual MINTEQ. It was assumed that heavy metals were adsorbed only onto the montmorillonite (main component of Na-bentonite). For the edge surface sites of montmorillonite two types of hydroxyl groups ($\equiv$EOH) with high and low affinity were involved in the sorption calculations. For modelling the binding of protons and heavy metal cations (Me^{2+}) onto the edge hydroxyl groups the diffuse double layer model (DDL) was used. The following protonation, deprotonation and complexation reactions for the edge ($\equiv$E) hydroxyl groups were considered (Bradbury and Baeyens, 1995):

$$\equiv EOH + H^+ \rightleftharpoons \equiv EOH_2^+ \qquad\qquad K_{(+)} \qquad\qquad (2.1)$$

$$\equiv EOH \rightleftharpoons \equiv EO^- + H^+ \qquad\qquad K_{(-)} \qquad\qquad (2.2)$$

$$\equiv EOH + Me^{2+} \rightleftharpoons \equiv EOMe^+ + H^+ \qquad\qquad K_{(Me)} \qquad\qquad (2.3)$$

$K_{(+)}$, $K_{(-)}$ and $K_{(Me)}$ are the intrinsic equilibrium constants for protonation, deprotonation and complexation reactions respectively and were calculated by using the mass action law, taking into account the electrostatic term (Bradbury and Baeyens, 1995):

$$\exp\left(\frac{\Delta ZF\,\Psi}{RT}\right) \qquad\qquad (2.4)$$

where ΔZ is the change in charge of the surface species, F is the Faraday constant, Ψ is the surface potential, R is the molar gas constant and T is the absolute temperature.

For the binding of the heavy metal cations onto the planar surface the fixed charge sites ($\equiv X^-$) with cation exchange potential were chosen:

$$\equiv XNa + 0.5Me^{2+} \rightleftharpoons \equiv XMe_{0.5} + Na^+ \qquad\qquad K^x_{(Me)} \qquad\qquad (2.5)$$

$K^x_{(Me)}$ is the equilibrium constant for cation exchange reaction.

The surface site capacities and the equilibrium constants used for the model calculations are given in Table 2.2.

The initial heavy metal concentration was set to 10 μmol/L and the concentration of montmorillonite to 200 and 500 mg/L. The sorption experiments showed that the concentrations of Na$^+$ cations in the supernatant were between 0.2 and 0.7 mmol/L. Therefore, the sodium cation and chloride concentrations were selected for the model calculations to be: $c(Na^+) = 0.5$ mmol/L and $c(Cl^-) = 0.5$ mmol/L. The model calculations were carried out for fixed pH values of 5 and 7 and for CO_2 equilibrium between aquatic and gas phase at 20°C and 1 atm.

Table 2.2. Surface site capacities and the equilibrium constants

	low affinity (edge site)	high affinity (edge site)	planar site
capacitiy of hydroxyl groups ($\equiv$EOH) in mmol/g	0.08[a]	0.002[a]	-
capacitiy of fixed charge sites ($\equiv$X$^-$) in mmol/g	-	-	0.87[a]
log $K_{(+)}$	5.4[b]	5.4[b]	-
log $K_{(-)}$	-6.7[b]	-6.7[b]	-
log $K_{(Cu)}$	0.6[c]	2.89[c]	-
log $K_{(Pb)}$	0.3[c]	4.65[c]	-
log $K_{(Zn)}$	-1.99[c]	0.99[c]	-
log $K^x_{(Cu)}$	-	-	2.8[d]
log $K^x_{(Pb)}$	-	-	2.8[d]
log $K^x_{(Zn)}$	-	-	2.7[d]

[a] Values from Bradbury and Baeyens (1995). [b] Values from Wanner et al. (1994). [c] Values from Parkhurst and Appelo (1999). [d] Values from Lothenbach et al. (1997)

2.3.4 Transport Experiments

The influence of Na-bentonite on the transport of heavy metal cations in porous media was investigated in the column experiments by using a "short pulse" laboratory column system with online metal and colloid detection. The eluent was pumped by a high-pressure liquid chromatography pump (HPLC pump 420, Kontron Instruments) into the glass column (length 171 mm, inner diameter 20 mm) filled with quartz sand in an upward flow. The flow rate was 1 mL/min. The deionised water was used as an eluent. For the injection of the samples an injection valve with an injection loop of 500 µL was used. The quartz sand was wet packed in the column. After packing, the column was equilibrated with an eluent for one week. This enabled the equilibration conditions in the column for the following experiments. The colloids were online detected at the column outlet by using of flow-through UV detector (Spectra System UV 1000, Spectra Physics Analytical). The UV signal was measured at $\lambda = 254$ nm. The UV detector allowed the detection of inorganic colloids by measure of light intensity attenuation through light scattering. For the detection of metals the column system was coupled with an inductively coupled plasma mass spectrometer (ICP-MS). For controlling the detector signal drift of the

ICP-MS system an internal standard containing 20 µg/L Rh, In and Ir each was continuously mixed to the eluent between the UV detector outlet and the ICP-MS inlet. A solution of sodium nitrate (Riedel-de Haën) with a concentration of 2 mmol/L was used as a conservative tracer for controlling the hydraulic conditions within the column and for the detection of the pore volume.

For the investigation of the colloid facilitated transport of heavy metals solutions of Cu, Pb and Zn in the presence and absence of Na-bentonite were injected into the column. These samples were prepared by using metal stock solutions and a Na-bentonite stock dispersion. The concentration of all metals was set to 10 µmol/L. The Na-bentonite in the samples had a concentration of 200 mg/L. The transport experiments were done at pH values of 5 and 7. The pH in the eluent and in the injection samples was adjusted by addition of sodium hydroxide or hydrochloric acid. Between injection series at pH 5 and pH 7, the column was rinsed by HNO_3 solution with a concentration of 0.1 mol/L followed by deionised water. After washing, the column was equilibrated with the concerning eluents for one week. This washing procedure enabled the release of adsorbed metals from the quartz sand surface and the regeneration of sorption places. By using the "short-pulse" technique it was possible to perform the transport experiments without significant blocking or filter ripening effects (Kretzschmar et al. 1997). For the calculation of recovery rates of colloids and metals bypass measurements were done. To avoid detector overflow during bypass injections, the samples were diluted to a ratio of 1:20. This dilution corresponds approximately to the dilution effect in the quartz sand column during transport experiments.

2.4 Results and Discussion

2.4.1 Influence of pH Value on the Aggregation of Na-bentonite

The stability of Na-bentonite was investigated in the titration experiments. Fig. 2.1 shows the zeta potential and particle size of Na-bentonite colloids as a function of the pH value. The zeta potential of bentonite at the original pH value (pH 7) was about -39 mV. The net zeta potential remained negative at the investigated pH range (pH 2 to pH 12) and as expected became less negative with decreasing pH value. It is attractive to assume that with decreasing pH value the surface of bentonite was more protonated. Decrease in the negative zeta potential can be caused also through the release of Al^{3+} from the bentonite structure and following readsorption onto the exchange sites, especially at the low pH values. At a pH value of 2, the zeta potential became more negative. Through the adsorption of anions (Cl⁻) and cations (Na⁺) at low and high pH values respectively the influence of ionic strength on the surface charge increased. Therefore, for the titration experiments without electrical conductivity adjustment the absolute value of the zeta potential at low and high pH values decreases (Müller 1996). The average particle size of bentonite remained stable (about 280 nm) in the pH range of 4 to 12. At the lower pH values (pH 4 to pH 2), the particle size increased rapidly and bentonite colloids agglomerated. At pH 2, bentonite agglomerates with average particle size of about 2 μm were detected.

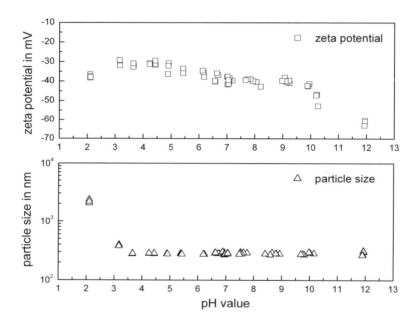

Fig. 2.1. Zeta potential and average particle size of Na-bentonite colloids as a function of pH value

Fig. 2.2 shows the particle size distribution of Na-bentonite colloids before (pH 7) and after (pH 2) titration. At the pH value of 2, the size of bentonite aggregates was partly above the detection limit of the zetasizer (6 μm). Similarly aggregation behaviour was detected for Na-montmorillonite particles by Rand et al. (1980). At low pH values, the lateral hydroxyl groups of bentonite are highly protonated and the edge sites have positive potential. It is well known that the face and edge surfaces of clay particles at low pH values are oppositely charged (Rand et al. 1980, Saunders et al 1999, Janek and Lagaly 2001, Tawari et al. 2001). Although the edge surface for montmorillonite particles is very small (<1%) compared to the total surface, the electrostatic attraction between the negatively charged face and positively charged edge sites probably favours the formation of agglomerates with a "house-of-cards" structure at extremely low pH values.

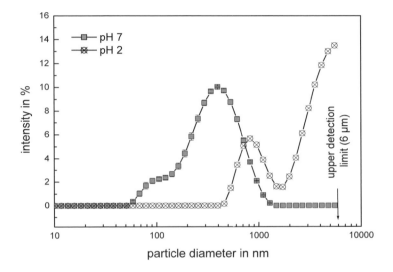

Fig. 2.2. Particle size distribution of Na-bentonite before (pH 7) and after (pH 2) titration (including standard derivation for pH 7)

2.4.2 Adsorption of Heavy Metals onto Na-bentonite

The experiments for the adsorption of Cu, Pb and Zn onto the Na-bentonite colloids were done at pH values of 5 and 7. The elimination of heavy metal ions from their solutions was very high (93% to 99%) for Na-bentonite concentrations of 200 mg/L and 500 mg/L (Table 2.3). In general, the metal adsorption capacities for the colloids were not exceeded in the concentration ranges investigated. It can be concluded that the available colloidal adsorption places were not the limiting factor in the column experiments (colloid concentration: 200 mg/L, metal concentration: 10 μmol/L).

Table 2.3. Adsorption of heavy metals (Me) onto Na-bentonite colloids as a function of bentonite concentration ($\rho_{bent.}$) and pH value, experimental (exp.) and calculated data

$\rho_{bent.}$ in mg/L	pH	metal	adsorbed metal in % (exp.)	adsorbed metal in % (calculated)			
				total	planar	edge low aff.	edge high aff.
200	5	Cu	93.3	99.55	98.66	0.18	0.71
		Pb	99	99.57	95.82	0.08	3.67
		Zn	97.1	99.27	99.25	0	0.02
	7	Cu	97.1	99.87	50.3	45.62	3.95
		Pb	99.4	99.83	55.79	40.05	3.99
		Zn	98.3	99.38	91.4	4.13	3.85
500	5	Cu	94.9	99.94	99.58	0.06	0.3
		Pb	98.4	99.95	93.92	0.03	6
		Zn	94.9	99.9	99.89	0	0.01
	7	Cu	98.5	99.98	43.35	47	9.63
		Pb	99.5	99.97	51.75	38.23	9.99
		Zn	99.2	99.91	90.42	1.58	7.91

The detected adsorption isotherms are presented on the Fig. 2.3. The isotherms show a small pH dependence of the adsorption of copper and lead onto the bentonite. At pH 7, a slightly higher sorption of copper and lead was observed then at pH 5. For zinc no significant influence of the pH value on the sorption onto the bentonite was found. Similar adsorption behaviours of different heavy metals on the montmorillonite were reported by Morton et al. (2001) and by Abollino et al. (2003). The heavy metal cations can adsorb on planar and edge sites of clay particles. At low background electrolyte concentrations, the adsorption on planar sites of montmorillonite is the predominant sorption mechanism. With increasing pH value the lateral hydroxyl groups of montmorillonite deprotonate, due to the increasing sorption of divalent metal cations on the edge sites. The edge/face ratio for montmorillonite particles is fairly small. Therefore the influence of the pH value on the total amount of adsorbed metals is low.

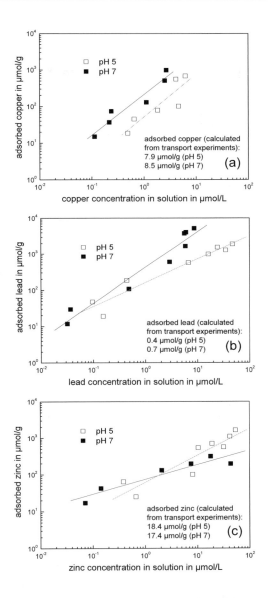

Fig. 2.3. Adsorption isotherms for Cu (a), Pb (b) and Zn (c) onto the Na-bentonite at pH values of 5 and 7

The results of the model calculations (Table 2.3) showed a high sorption of heavy metal cations onto planar sites for the pH value of 5. At pH 7 the binding of heavy metal cations on the edge site hydroxyl groups increased. This effect was stronger especially for copper and lead. Zinc cations were predominantly adsorbed onto the planar sites at both pH values. Due to the competing sorption processes, at pH 7 heavy metal cations were partly displaced from the planar sites to the edge binding sites. The initial heavy metal concentration was chosen similar to the concentrations in natural systems. It is obvious that this concentration (10 µmol/L) was much lower than the total capacity of the surface groups (190 µmol/L for a bentonite concentration of 200 mg/L and 476 µmol/L for a bentonite concentration of 500 mg/L). The calculated total amount of adsorbed heavy metals was very high (about 99%). This is in good agreement with the experimental data.

2.4.3 Influence of Na-bentonite on the Transport of Heavy Metals

In the transport experiments, the breakthrough behaviour of heavy metals and Na-bentonite colloids in porous media using the "short pulse" laboratory column system with online metal and colloid detection was investigated. The Figs. 2.4, 2.5 and 2.6 represent the breakthrough curves of the heavy metals and colloids at pH values of 5 and 7.

The breakthrough curves of Na-bentonite were detected as UV signal using a UV detector and as Si, Mg and Al signals using a ICP-MS system. These elements can be used for the detection of Na-bentonite because they are its major components (see Table 2.1). The breakthrough curves show that the correlation between UV, Si, Mg and Al signals is very good.

In most cases Na-bentonite colloids elute earlier (0.93 - 0.94 pore volumes) than the tracer anions (NO_3^-, 1.0 pore volume). This effect is similar to the size exclusion principle in chromatography. During the transport through the porous media, bentonite particles are using only the large pores available in the quartz sand packing. In opposite to this, the water molecules or dissolved ions, which are much smaller compared to the bentonite particles (280 nm), can diffuse into the small pores of the quartz sand packing and by this arrive later at the column end than the colloids. Several authors have reported the influence of the size exclusion effect on the transport of colloids in porous media (Harvey et al 1989, Kretzschmar and Sticher 1997, Kretzschmar et al. 1997, Huber et al. 2000, Metreveli et al. 2005).

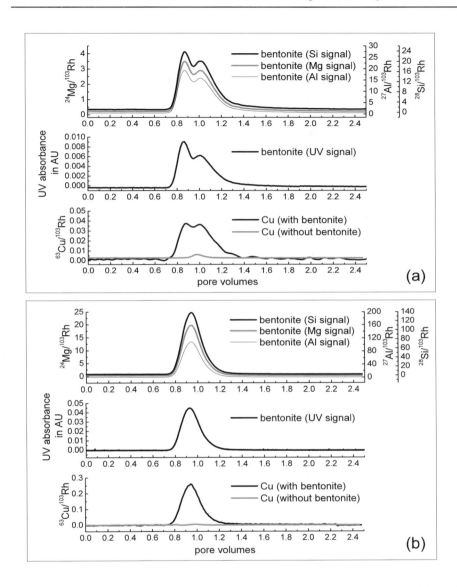

Fig. 2.4. Breakthrough curves for copper and Na-bentonite at pH 5 (a) and pH 7 (b)

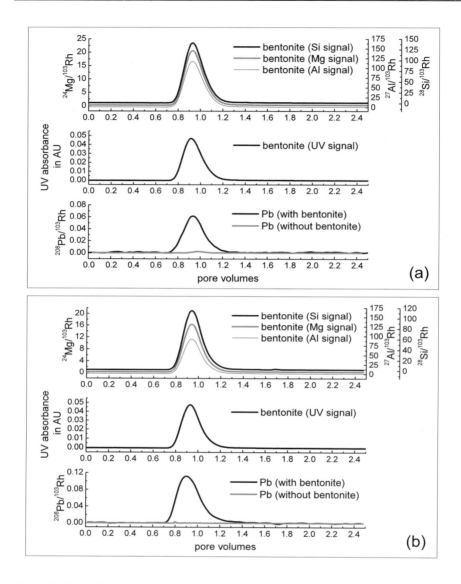

Fig. 2.5. Breakthrough curves for lead and Na-bentonite at pH 5 (a) and pH 7 (b)

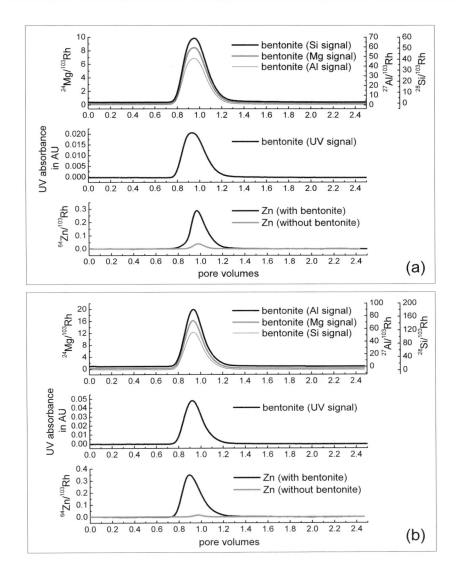

Fig. 2.6. Breakthrough curves for zinc and Na-bentonite at pH 5 (a) and pH 7 (b)

In the presence of copper at a pH value of 5 bentonite colloids elute in two fractions (Fig. 2.4, (a)), the first one at 0.86 and the second one at 1.01 pore volumes. Adsorption of copper on the bentonite favours probably the formation of bentonite aggregates (first fraction), which elute according to the size exclusion effect earlier than the primer bentonite particles. The other reason for the splitting of the breakthrough curves may be the increasing interaction between the quartz sand surface and the bentonite particles at low pH values in the presence of bivalent metal cations. At low pH values, the surface of quartz sand and bentonite colloids is more protonated then at high pH values and the surface potential is less negative. In addition the adsorption of heavy metals causes a further decrease of the negative surface potential. As a consequence of the decrease in negative surface potential the attractive forces between bentonite particles and quartz grains increases and the elution of bentonite colloids decelerates. The small tailing of the Si, Mg, Al and UV signals (Fig. 2.4, (a)) indicate the attraction between bentonite colloids and column packing material. The splitting of the breakthrough curves was only observed for bentonite colloids in the presence of copper at a pH value of 5. At the injection of bentonite colloids in the presence of lead and zinc only one bentonite fraction was found at the column outlet (Figs. 2.5 and 2.6).

For the transport experiments recovery calculations were done. The recoveries for Na-bentonite colloids were calculated from the Mg, Si, Al and UV signals. The results of the recovery calculations are presented in Fig. 2.7. The recoveries for Na-bentonite calculated from different signals and detectors (ICP-MS and UV detector) correlate well with each other. The results show that the coupling method (column/UV/ICP-MS) can be successfully used for the qualitative and quantitative characterisation of the transport of colloids and metals through porous media. Furthermore the calculated recoveries of Na-bentonite confirm the increasing interaction between bentonite colloids and quartz grains with decreasing pH value. At pH 7, the recoveries of Na-bentonite are high (between 83% - 93%). At the high pH values, the surfaces of the quartz sand and the bentonite colloids are more deprotonated and the surface potentials become more negative. Due to the increase of the repulsion forces the colloids remain highly mobile. At a pH value of 5, the surfaces of the colloids and the quartz grains are more protonated. The increase of the attraction forces results in an increase of particle deposition on the quartz sand surface. However, the deposition rates for bentonite particles at pH 5 were different in the presence of different heavy metals. So showed the copper ions a stronger influence on the particle deposition than the zinc and lead ions. In the presence of copper only about 20% of the injected bentonite colloids were found at the column outlet. In the presence of zinc and lead ions the recoveries of bentonite were about 42% and 80% respectively.

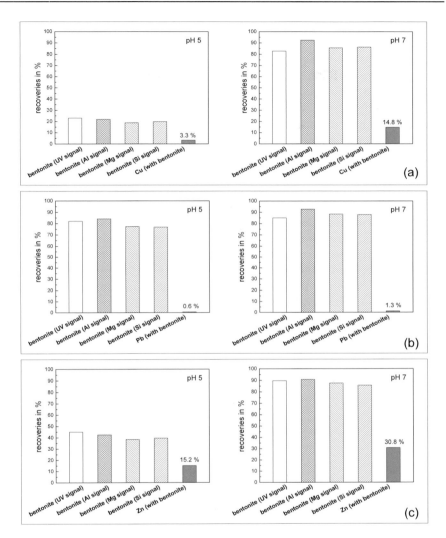

Fig. 2.7. Recoveries of Na-bentonite and heavy metals calculated from column and bypass measurements: recoveries at pH values of 5 and 7 for Na-bentonite and the metals copper (a), lead (b), and zinc (c)

In the transport experiments with colloid free metal solutions, no significant transport of heavy metal ions was found. At both pH values (pH 5 and pH 7), in the absence of bentonite colloids the heavy metals were totally adsorbed on the quartz sand surface (Figs. 2.4, 2.5 and 2.6). This is in good agreement with the results of transport experiments reported by Schmitt et al. (2003). By the injections of bentonite containing heavy metal solutions a colloid facilitated transport for all heavy metals was observed. The breakthrough curves show that the elution volume of the heavy metals correlates well with the elution volume of the bentonite colloids. It is interesting to note that the heavy metals showed different transport behaviour. Recoveries of heavy metals calculated from transport experiments were low, generally between 0.6% and 30.8% for the column investigated. The eluted amount of metals increased in the order: Pb < Cu < Zn (Fig. 2.7). As detected in the sorption experiments, about 93% - 99% of the heavy metals were adsorbed onto the bentonite (Table 2.3). From the recovery rates the amount of adsorbed heavy metals onto bentonite at the column outlet was calculated. For these calculations it was assumed that heavy metals, detected at the column outlet were only attached to the colloids. The results of these calculations are shown in Fig. 2.3. The adsorbed amounts for all heavy metals were lower then the values detected in the sorption experiments. During the passage through the column, the competitive sorption processes between bentonite and the column packing material obviously result in a desorption of heavy metals from the bentonite colloids and an immobilisation on the quartz sand surface.

Table 2.4. Recovery ratios of heavy metals and bentonite (revealed by different properties)

	pH 5	pH 7
Cu/UV	0.144	0.179
Cu/Al	0.151	0.160
Cu/Mg	0.176	0.173
Cu/Si	0.167	0.171
Pb/UV	0.007	0.015
Pb/Al	0.007	0.014
Pb/Mg	0.008	0.014
Pb/Si	0.008	0.014
Zn/UV	0.338	0.343
Zn/Al	0.358	0.339
Zn/Mg	0.395	0.351
Zn/Si	0.385	0.359

Furthermore there was an influence of the pH value on the colloid facilitated transport of heavy metals. At pH 7, the amount of transported heavy metals was higher than at pH 5 (Fig. 2.7). This effect is due to the increasing deposition of bentonite on the quartz grains with decreasing pH value. For both pH values the (heavy metal)/bentonite recovery ratios were calculated (Table 2.4). They were for both pH values approximately the same. This confirms that the pH value has no significant effect on the sorption of heavy metals on the Na-bentonite, and that the transport of heavy metals through the column is mainly influenced by the particle deposition and competitive sorption processes between bentonite and the column packing material.

2.5 Conclusions

The titration experiments showed that with decreasing pH value the net zeta potential of Na-bentonite colloids became less negative. At the pH range of 4 to 12, Na-bentonite colloids remained stable. At the lower pH values (pH 2 to 4), Na-bentonite particles aggregated rapidly. In the sorption experiments, a small pH dependence of the adsorption of copper and lead onto the Na-bentonite could be detected. For zinc no significant influence of the pH value on the sorption behaviour was found. The results show that laboratory experiments performed under well defined conditions are well suited to characterize the transport relevant properties of colloidal surfaces.

The column experiments with online metal and colloid detection resulted in a good correlation between different signals of different detectors (UV, ICP-MS). The Na-bentonite particles were highly mobile at a pH value of 7. At pH 5, the deposition of Na-bentonite colloids on the column packing material (quartz sand) could be observed. The deposition of Na-bentonite particles was different in the presence of different heavy metals and increased in the order: Pb < Zn < Cu. In the absence of Na-bentonite colloids at both pH values (pH 5 and 7) no significant transport of heavy metal ions was found. The injections of the Na-bentonite containing heavy metal solutions led for all heavy metals investigated to a colloid facilitated transport. The amount of transported heavy metals in the presence of Na-bentonite colloids was quite different for the individual metals investigated and turned out to be mainly influenced by the particle deposition and competitive sorption processes between Na-bentonite and the column packing material.

The experiments demonstrate clearly to power of column experiments with coupled advanced detection methods. The results obtained are meaningful for the understanding of the transport processes in water saturated porous media. The principles of size exclusion chromatography can be applied to a far extend also to the separation and transport processes of colloids in porous media. Model calculations based on a multi site assumption for the colloidal surface showed a good agreement with experimental data. This is promising for a reliable prediction of transport phenomena even in complex systems like natural and polluted aquifers.

Acknowledgement

The authors thank the German research council (DFG) for financial support within the KORESI project (colloidal transport of substances by seepage of rainwater).

References

Abollino O, Aceto M, Malandrino M, Sarzanini C, Mentasti E (2003) Adsorption of heavy metals on Na-montmorillonite. Effect of pH and organic substances. Water Research 37:1619-1627
Bunn RA, Magelky RD, Ryan JN, Elimelech M (2002) Mobilization of natural colloids from an iron oxide-coated sand aquifer: effect of pH and ionic strength. Environ Sci Technol 36:314-322
Bradbury MH, Baeyens B (1995) A Quantitative mechanistic description of Ni, Zn and Ca sorption on Na-montmorillonite. Part III: modeling. PSI Bericht Nr. 95-12, Paul Scherrer Institut, Villigen, Switzerland
Christl I, Kretzschmar R (2001) Interaction of copper and fulvic acid at the hematite-water interface. Geochimica et Cosmochimica Acta 65, 20:3435-3442
DIN 38 409-H1-1, Teil 1 (1987) Deutsche Einheitsverfahren zur Wasser-, Abwasser- und Schlammuntersuchung; Summarische Wirkungs- und Stoffkenngrößen (Gruppe H); Bestimmung des Gesamttrockenrückstandes, des Filtrattrockenrückstandes und des Glührückstandes (H 1)
Elimelech M, O`Melia CR (1990) Kinetics of deposition of colloidal particles in porous media. Environ Sci Technol 24:1528-1536.
Flury M, Mathison JB, Harsh JB (2002) In situ mobilization of colloids and transport of cesium in Hanford sediments. Environ Sci Technol 36:5335-5341
Grolimund D, Barmettler K, Borkovec M (2001) Release and transport of colloidal particles in natural porous media. 2. Experimental results and effects of ligands. Water Resour Research 37, 3:571-582

Grolimund D, Borkovec M, Barmettler K, Sticher H (1996) Colloid-facilitated transport of strongly sorbing contaminants in natural porous media: a laboratory column study. Environ Sci Technol 30:3118-3123

Grolimund D, Elimelech M, Borkovec M, Barmettler K, Kretzschmar R, Sticher H (1998) Transport of in situ mobilized colloidal particles in packed soil columns. Environ Sci Technol 32:3562-3569

Harvey RW, George LH, Smith RL, LeBlanc DR (1989) Transport of microspheres and indigenous bacteria through a sandy aquifer: results of natural- and forced-gradient tracer experiments. Environ Sci Technol 23:51-56

Huber N, Baumann T, Niessner R (2000) Assessment of colloid filtration in natural porous media by filtration theory. Environ Sci Technol 34: 3774-3779

Janek M, Lagaly G (2001) Proton saturation and rheological properties of smectite dispersions. Applied Clay Science 19:121-130

Johnson PR, Elimelech M (1995) Dynamics of colloid deposition in porous media: blocking based on random sequential adsorption. Langmuir 11:801-812

Johnson PR, Sun N, Elimelech M (1996) Colloid transport in geochemically heterogeneous porous media: modeling and measurements. Environ Sci Technol 30:3284-3293

Kretzschmar R, Barmettler K, Grolimund D, Yan Y, Borkovec M, Sticher H (1997) Experimental determination of colloid deposition rates and collision efficiencies in natural porous media. Water Resour Research 33, 5:1129-1137

Kretzschmar R, Borkovec M, Grolimund D, Elimelech M (1999) Mobile subsurface colloids and their role in contaminant transport. Advances in Agronomy 66:121-193

Kretzschmar R, Robarge WP, Amoozegar A (1995) Influence of natural organic matter on colloid transport through saprolite. Water Resour Research 31, 3:435-445

Kretzschmar R, Sticher H (1997) Transport of humic-coated iron oxide colloids in a sandy soil: influence of Ca^{2+} and trace metals. Environ Sci Technol 31:3497-3504

Kulkarni P, Sureshkumar R, Biswas P (2005) Hierarchical approach to model multilayer colloidal deposition in porous media. Environ Sci Technol 39:6361-6370

Litton GM, Olson TM (1993) Colloid deposition rates on silica bed media and artifacts related to collector surface preparation methods. Environ Sci Technol 27:185-193

Liu D, Johnson PR, Elimelech M (1995) Colloid deposition dynamics in flow through porous media: role of electrolyte concentration. Environ Sci Technol 29:2963-2973

Lothenbach B, Furrer G, Schulin R (1997) Immobilization of heavy metals by polynuclear aluminium and montmorillonite compounds. Environ Sci Technol 31:1452-1462

McCarthy JF, Zachara JM (1989) Subsurface transport of contaminants. Environ Sci Technol 23:496-502

Metreveli G, Kaulisch E-M, Frimmel FH (2005) Coupling of a column system with ICP-MS for the characterisation of colloid-mediated metal(loid) transport in porous media. Acta hydrochim hydrobiol 33, 4:337-345

Morton JD, Semrau JD, Hayes KF (2001) An X-ray absorption spectroscopy study of the structure and reversibility of copper adsorbed to montmorillonite clay. Geochimica et Cosmochimica Acta 65, 16:2709-2722

Müller RH (1996) Zetapotential und Partikelladung in der Laborpraxis: Einführung in die Theorie, praktische Meßdurchführung, Dateninterpretation. Wissenschaftliche Verlagsgesellschaft, Stuttgart

Oste LA, Temminghoff EJM, Van Riemsdijk WH (2002) Solid-solution partitioning of organic matter in soils as influenced by an increase in pH or Ca concentration. Environ Sci Technol 36:208-214

Parkhurst DL, Appelo CAJ (1999) User's guide to phreeqc (version 2) - a computer program for speciation, batch-reaction, one-dimensional transport, and inverse geochemical calculations. Water-Resources Investigation Report 99-4259, U.S. Geological Survey, Denver, Colorado

Rand B, Pekenć E, Goodwin JW, Smith RW (1980) Investigation into the existence of edge-face coagulated structures in Na-montmorillonite suspensions. J C S Faraday I 76:225-235

Roy SB, Dzombak DA (1996) Colloid release and transport processes in natural porous media. Colloids and Surfaces, A: Physicochemical and Engineering Aspects 107:245-262

Roy SB, Dzombak DA (1997) Chemical factors influencing colloid-facilitated transport of contaminants in porous media. Environ Sci Technol 31: 656-664

Ryan JN, Elimelech M (1996) Colloid mobilization and transport in groundwater. Colloids and Surfaces, A: Physicochemical and Engineering Aspects 107:1-56

Ryde N, Kallay N, Matijević E (1991) Particle adhesion in model systems. Part 14. - Experimental evaluation of multilayer deposition. J Chem Soc Faraday Trans 87, 9:1377-1381

Saiers JE, Hornberger GM (1999) The influence of ionic strength on the facilitated transport of cesium by kaolinite colloids. Water Resour Research 35, 6:1713-1727

Saunders JM, Goodwin JW, Richardson RM, Vincent B (1999) A small-angle x-ray scattering study of the structure of aqueous laponite dispersions. J Phys Chem B 103:9211-9218

Schmitt D, Saravia F, Frimmel FH, Schuessler W (2003) NOM-facilitated transport of metal ions in aquifers: importance of complex-dissociation kinetics and colloid formation. Water Research 37:3541-3550

Tawari SL, Koch DL, Cohen C (2001) Electrical double-layer effects on the Brownian diffusivity and aggregation rate of laponite clay particles. Journal of Colloid and Interface Science 240:54-66

Vinten AJA, Yaron B, Nye PH (1983) Vertical transport of pesticides into soil when adsorbed on suspended particles. J Agric Food Chem 31:662-664

Wanner H, Albinsson Y, Karnland O, Wieland E, Wersin P, Charlet L (1994) The acid/base chemistry of montmorillonite. Radiochimica Acta 66, 67:157-162

Weng L, Fest EPMJ, Fillius J, Temminghoff EJM, Van Riemsdijk WH (2002) Transport of humic and fulvic acids in relation to metal mobility in a copper-contaminated acid sandy soil. Environ Sci Technol 36:1699-1704

Zhuang J, Flury M, Jin Y (2003) Colloid-facilitated Cs transport through water-saturated Hanford sediment and Ottawa sand. Environ Sci Technol 37:4905-4911

3 Colloid Transport Processes: Experimental Evidence from the Pore Scale to the Field Scale

Thomas Baumann

Institute of Hydrochemistry, TU Munich, Germany

3.1 Abstract

Aquatic colloids play a central role in mediating the transport of contaminants in subsurface media. Depending on the environmental conditions, the transport of contaminants can be enhanced or retarded in the presence of colloids. The changes of the transport velocities and mass transfer rates of colloid-associated contaminants are of high environmental relevance.

The keys to the assessment of the effects of colloids on subsurface transport is the colloid-contaminant interaction and the mobility of the colloid itself. While the theory of colloid transport was laid out in the late 1950's, the actual parametrization of colloid transport has long been concealed by the used experimental conditions. Classical column tests obscure the pore scale processes by limiting the access to pore scale heterogeneities. Recently the transport of colloids and contaminants was unveiled using magnetic resonance imaging (MRI) and micromodel experiments.

MRI provides non-invasive access to dynamic processes in porous and fractured media with a spatial resolution down to $200{\times}200{\times}200\ \mu m^3$ and a temporal resolution in the seconds range. MRI data were used to characterize the pore space, to measure the flow and diffusion of water, to quantify the attachment of colloids and the transport of contaminants in porous media. The transport of water, certain metal ions and several colloids can be measured without additional tracers. Other potential contaminants are accessible using magnetic tagging, or indirectly by their effect on the relaxation times of water. The main limitations of MRI include susceptibility artefacts and the dependence of the MRI signal on a variety of boundary conditions, sometimes leading to ambiguous calibration curves.

Micromodels, i.e. porous structures etched in silicon wafers, glass, or other materials, provide single-particle–single-pore access. With a spatial resolution below 1 μm and a time resolution in the millisecond range, the processes at the pore scale are accessible. The results demonstrate that the theoretical framework of colloid filtration can be applied to the pore scale transport of colloids. It was shown, that the main causes for deviations of experimental results from theoretical predictions are particle-particle interactions, inaccurate descriptions of the flow field, chemically heterogeneous surfaces and additional collector surfaces. Air bubbles and other non-miscible liquids can be considered as temporary collectors, since the colloids attached to the air-waterinterface are released when the air bubble dissolves. Non-polar interfaces cause a repartitioning of non-polar contaminants, thus limiting the effects of colloid transport.

While MRI and micromodels can significantly enhance our understanding of colloid transport phenomena, upscaling to field conditions is still difficult and requires an extensive and detailed data set. Laboratory, pilot scale and field investigations provide a quick way to obtain site and situation specific data on colloid transport processes. In column tests, the attachment probability, summarizing particle-collector and to some extent particle-particle interactions, can be parametrized using the ionic strength and the dominating cation of the solution. Column tests illustrated that the transport of colloids is scale dependent. Since the typical length scale of column tests is well below the theoretical transport distance, column tests tend to overpredict colloid transport. Changes of the hydrodynamic and hydrochemical conditions can cause significant colloid mobilization. This effect was also observed at the pilot scale and at the field scale.

Pilot and field scale experiments showed that the transport of metal ions is only very little affected in the presence of colloids. At transport distances of 200 m in a calcareous gravel aquifer, about 10% of the metal ions were found associated to colloids. Since the transport distance of the colloids themselves is far less, a dynamic equilibrium between colloids and dissolved metal ions was proposed. With increasing transport distance, a slight increase of colloid-associated metal ions has been found. The results limit the importance of colloid transport in the investigated shallow aquifers to otherwise immobile contaminants like PAHs or radionuclides.

One scenario, where several of the preconditions for colloid transport seemed to be fulfilled, is the emission of contaminants from landfills with direct contact to groundwater. Here, a high concentration of colloids and high concentrations of contaminants come together. Our research shows, that the interface between the landfill body and the groundwater is an effective hydrochemical barrier for colloids. The colloids downgradient of the disposal sites and the colloids inside the disposal differ significantly. We propose a self-sealing effect of the landfill, caused by landfill colloids being filtered at the interface, and by this reducing the hydraulic conductivity of the interface. Field investigations verified this hypothesis. Moreover, detailed investigations of the colloid-metal associations revealed that the highest concentrations of metal ions were found with particles greater than 10 μm, i.e. the least mobile particles.

3.2 Pore Scale Phenomena

3.2.1 Visualization and Quantification of Colloid Transport on the Pore Scale: The Micromodel-Approach

The description of colloid transport phenomena suffers from a principal shortcoming: the theory was developed for a single particle–single collector system (Happel, 1958) and later extended to a single particle–multi collector system (Rajagopalan and Tien, 1976; Rajagopalan and Chu, 1982). In theory, the collector and the fluid flow around the collector are well defined. The theoretical framework is, however, usually applied to systems where the only variable which is accessible is the average flow velocity.

The boundary conditions, especially the pore topology, the local flow velocities, and the chemical heterogeneity of the surface, can vary in the pore space. In column tests, the packing density and the topology of the pore network are unknown and local physical and chemical heterogeneities cannot be assessed (Sugita and Gillham, 1995). Also, preferential flow inside the column is obscured. Therefore, the inverse derivation of filtration parameters, e.g. the attachment efficiency, are representative only for the length scale of the column. As a result, upscaling and downscaling becomes difficult and experimentally derived parameters do not always conform to theory.

The shortcomings in the description of the boundary conditions have led to empirical extensions of filtration theory (Tufenkji and Elimelech, 2004). While these improve the performance of filtration theory to predict colloid attachment on the bench scale, the principal problem remains: What happens at the pore scale? Finally, filtration theory cannot be applied to multiphase systems (Gao et al., 2004; Saiers and Lenhart, 2003a,b). Here additional data is needed to develop a theoretical framework for colloid transport.

To answer this set of questions, analytical techniques are required which allow single particle tracking. In groundwater and vadose zone studies, micromodels are representations of porous media etched into silicon wafers, glass, or polymers (Soll et al., 1993). Sometimes thin layers of glass beads, or sand, embedded between glass plates are also referred to as micromodels. The main purpose of micromodel experiments is to increase spatial and temporal resolution, and to provide direct quantitative access to processes at the pore scale. Micromodel experiments with colloids have been described by (Wan and Wilson, 1994b; Wan et al., 1996). In one study, the role of the gas-water interface for the transport of fluorescent latex beads, clay particles and bacteria was investigated in an etched glass micromodel. The gas-water interface turned out to be an effective sink for both hydrophobic and hydrophilic particles (Wan and Wilson, 1994b). The data was validated with column tests (Wan and Wilson, 1994a). In the other study, fluorescent colloids were used as a tracer for the flow velocities in a glass micromodel with a pore network divided by a fracture with an aperture of 1mm. The local velocity in the fracture was then measured using recorded video data and compared to the theoretical parabolic flow profile and showed good agreement (Wan et al., 1996).

With the availability of advanced image acquisition and numerical modelling techniques, micromodel experiments came into scientific focus again. Current approaches combine the power of direct observation in micromodels, with a pore-scale model analysis of water flow paths and velocities. In contrast to previous experimental studies of colloids using micromodels, we used a plasma etching technique to create highly controlled 2D pore structures with vertical walls and flat bottom (Fig. 3.1). This allows us to quantitatively compare colloid flow paths and velocities in pore bodies to those determined for water using the Lattice-Boltzmann model (Baumann and Werth, 2004).

Two different sizes of spherical, fluorescent colloids, 691 nm and 1960 nm (Duke Scientific, Palo Alto, CA) were used. They were surfactant free, and the reported particle density is 1.05 g/cm^3. The input pulse concentrations for the 691 nm and 1960 nm colloids were $1.1 \cdot 10^{12}$/L and $4.8 \cdot 10^{10}$/L, respectively, the input mass concentration was 0.2 mg/L for both types of colloids. The zeta potential of the colloids was measured using a Zetaphoremeter III (SEPHY, Paris, F). The values in deionized distilled water (DDI water) at a pH of 5.6 (unbuffered dispersion) were -50.2 ± 4.6 mV, and -26.0 ± 2.8 mV for the 691 nm and 1960 nm colloids, respectively. At pH 5.6 the surface of the micromodel also is negatively charged.

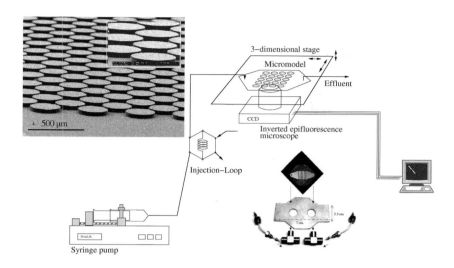

Fig. 3.1. Experimental Setup and SEM picture of the homogeneous micromodel (Baumann and Werth 2004, Copyright with permission from Soil Science Society of America).

A composite order image was constructed by overlaying four images taken 2 seconds apart, where colloids in each image were assigned a different color. The composite order image of 691 nm colloids at 15 μL/h in the micromodel pore network is shown in Fig. 3.2. Results at other flow rates and for 1960 nm colloids were similar and are not shown. Flow is from left to right. The exposure time for each image was 500 ms and 100× magnification was used (i.e. 10× objective).

Colloids appear as dots and streaks in the image. Some dots and streaks are not in the focal plane, i.e. not in the vertical center of the micromodel, and these appear blurry (e.g. **c**). Faster moving colloids appear as streaks, where the length of each streak is proportional to the colloid velocity. The longest streaks are in pore throats, the shortest near grain walls. The flow velocities of the particles in the pore throats reach about 1000 cm/day, which is five times higher than the average linear velocity and in agreement with the calculated 2D velocity distribution.

Near grain walls there are cases where colloids appear as streaks (cases **a** & **b**) and as dots (cases **c** & **d**). When colloids appear as dots, they did not appreciably move during the 500 ms exposure time. For these cases, the composite order image illustrates that colloids moved between images taken 2 seconds apart. For example, colloids at position **c** and **d** in the image appear as dots along the grain wall; successive images show that these colloids move only a few colloid diameters every 2 seconds.

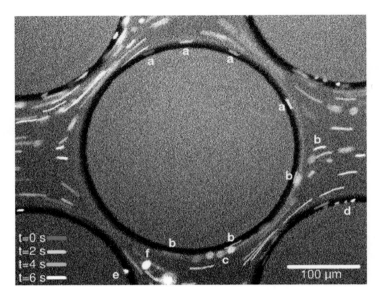

Fig. 3.2. A composite order image of 691 nm colloids in the homogeneous micromodel. Flow is from left to right at 15 µL/h. The color code indicates the position of the colloids at time steps 2 s apart (Baumann and Werth 2004, Copyright with permission from Soil Science Society of America).

In some cases streaks appear to be closer to the grain surface than dots. This seems counter-intuitive since colloids closer to grain walls should move slower. However, the micromodel is not a two-dimensional flow system. Flow across the depth of a pore channel is parabolic. As a result, the water velocity near the top and bottom of a pore channel is slower than at the mid-height.

The colloid at position **c** is out of focus, i.e. near the top or bottom of the flow channel, whereas, the colloid at position **b** is in focus, i.e. in the vertical center of the micromodel. This explains why colloid **b** moves faster than colloid **c**. Colloid **d** is also in focus and it is closer to the grain surface than colloid **b**. As a result, it moves slower than **b**. From Fig. 3.2 it can also be seen, that one colloid close to a grain surface (position **e**) is not moving at all. This colloid seems to be attached to the surface. Another colloid (position **f**) seems to be attached to the top or bottom of the micromodel.

Two dimensional (2D) water flow in the micromodel pores was previously simulated with the Lattice-Boltzmann (LB) method (Knutson et al., 2001). Images of the 1960 nm colloids are shown in Fig. 3.3 after breakthrough, i.e. the concentration of the colloids equals the input concentration, at 100× magnification. Also shown are calculated flow lines based on the Lattice-Boltzman model (Fig. 3.3b). In all cases, flow is from left to right at a flow rate of 150 µL/h. As in Fig. 3.2, the longest streaks are near the middle of pore throats and the shortest near grain walls. In addition, for a given streak, the intensity is less in a pore throat than in a pore body, indicating greater flow velocities in pore throats. The intensity also varies between streaks. However, it is not possible to correlate this with colloid velocity because colloids at different vertical positions (i.e. colloids that are in or out of focus) have different intensities.

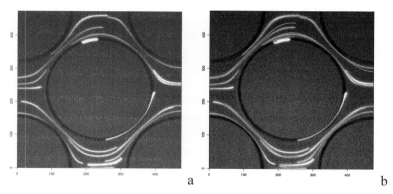

a   b

Fig. 3.3. Measured colloid trajectories (a) and an overlay of calculated and measured trajectories (b) for the 1960 nm colloids in the homogeneous micromodel during breakthrough. Flow is from left to right. Exposure time is 500 ms (Baumann and Werth 2004, Copyright with permission from Soil Science Society of America).

Calculated flow paths in Fig. 3.3b coincide well with measured streak lengths. A total of 56 individual colloid streaks from the images in Fig. 3.2 and 3.3, and from an image of the 1960 nm colloids at 150 µL/h at an earlier time during breakthrough (i.e. when the concentration was lower than in Fig. 3.3) were compared to model results. For about 50% of the colloids, measured streak lengths and locations matched the calculated trajectories very well (i.e. the length of measured and calculated trajectories was within ±10%, the lateral deviation was within ±2.5 µm). This is near the maximum of what could have been expected with the 2D approximation of the 3D flow field. The result indicates that in many cases colloids move along with the calculated flow of water in the pore structure and the model does a good job of determining colloid flow paths and velocities.

For about 20% of the colloids, measured streak locations matched the calculated trajectories, but the measured streak lengths were shorter (i.e. characterized by a smaller velocity). Most of these colloids were out of focus, i.e. the colloids are moving at the top or at the bottom of the flow channel; these colloids were likely affected by slower flow at the top and bottom of flow channels. As previously noted, this is a limitation of applying the 2D model to the 3D micromodel and subject for future research. On the other hand, the velocities of colloids in heterogeneous pore networks can be used to measure the velocity distribution which is otherwise not yet accessible due to limited computing capacities.

While the described experiments are useful for understanding and quantifying the filtration efficiency under fully saturated conditions, environmental systems are usually more complex. There is an ongoing discussion whether the air-water-interface, or the air-water-solid-interface is more important in colloid retention. The discussion started with the first micromodel experiments (Wan and Wilson, 1994b). Newer data suggests that colloids are mainly retained at the air-water-solid-interface (Crist et al., 2005, 2004). These findings are under controversial discussion.

In any case, experiments with homogeneous and heterogeneous micromodels show, that the retention of colloids at the air-water-(solid-)interface is temporary. At first the colloids are loosely attached to the gas-bubble and moving around the bubbles circumference. As soon as the gas bubble dissolves, the colloids are released into the water flow (Baumann et al., 2002b). In this sense, the air-water-interface presents a non-permanent collector (Saiers et al., 2003).

Colloid transport in more complex multiphase systems has not yet been addressed in micromodel experiments. Under environmental conditions, nonaqueous- phase liquids (NAPLs) can be present in the subsurface (Sahloul et al., 2002; Jia et al., 1999). The NAPLs can be considered as another hydrophobic interface, thus colloids are most likely retained at the water-NAPL-interface. This is shown in Fig. 3.4. In contrast to the air-water-interface, which is non-reactive, partitioning effects are occurring at the water-NAPL-interface. In this experiment, the polystyrol colloids attached to the n-octanol bubble slowly disintegrated and the tracer dye partitioned into the n-octanol (Baumann and Niessner, 2006).

While the kinetics of these processes are not yet quantified, the experimental data provides strong evidence for a mass transfer of contaminants between colloids and NAPLs. This has strong implications on the assessment of colloidal transport phenomena in contaminated areas.

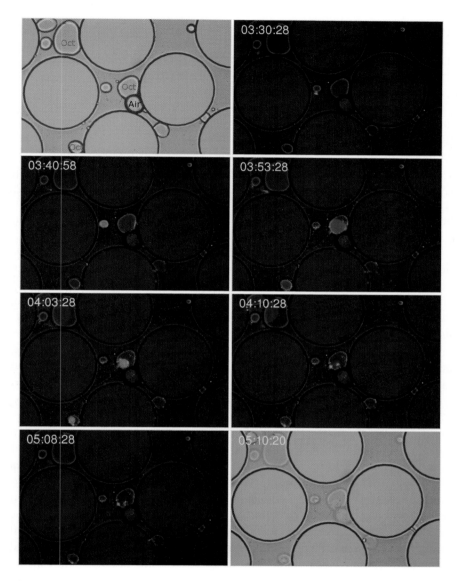

Fig. 3.4. Colloids in a water-air-n-octanol system. The diameter of the cylinders is 300 μm. Flow is from left to right at 50 μL/h (modified from Baumann and Niessner 2006).

3.3 Column Scale Phenomena

3.3.1 Visualization and Quantification of Colloid Transport on the Column Scale: Magnetic Resonance Imaging

Conventional methods to investigate colloid transport often involve column studies, where colloid concentrations are measured at the column effluent or at selected points along the column length. Unfortunately, such methods do not clearly distinguish how spatial and temporal changes in hydrochemical and hydrodynamic conditions affect colloid transport. For example, breakthrough curves (BTCs) obtained from column effluent represent some average behavior of colloids in the column. Since different heterogeneous realizations can contribute to such BTCs, the processes that control colloid transport in the column are obscured.

Conventional methods also have the potential to disrupt *in situ* conditions (Weisbrod et al., 1996; Backhus et al., 1993). For example, large volume or quickly withdrawn samples can disrupt the local flow field or alter column chemistry, thereby yielding results that do not represent those in undisturbed columns. Consequently, non-invasive *in situ* and fast detection techniques are required to resolve colloid transport.

Promising non-invasive detection techniques include γ-ray tomography, positron emission tomography (PET), x-ray tomography, and synchrotron radiation. γ-ray and PET are sensitive to radioactivity. Colloids tagged with radioisotopes can be imaged at resolutions approaching $5 \times 5 \times 5 \ mm^3$ (Greswell et al., 1998). The acquisition time is on the order of 30 min/picture. X-ray tomography is sensitive to changes of density in a system. The acquisition times for x-ray tomography are in the range of minutes per image. For colloids that are denser than water, static images can be obtained at resolutions approaching 40 μm. With synchrotron radiation, colloids in aqueous solution can be characterized *in situ* with a spatial resolution below 100 nm. Chemical properties of the colloids can also be measured (Megens et al., 1997; Neuhäusler et al., 1999). The typical thickness of a synchrotron sample for colloid detection should not exceed some μm, which makes this technique unsuitable for column studies. The acquisition time for synchrotron images is slightly longer than that for x-ray tomography.

Another promising non-invasive detection technique is magnetic resonance imaging (MRI). MRI is sensitive to changes in the local magnetic field. The local magnetic field can be altered in a controlled way by selective excitation of magnetically susceptible nuclei. MRI is robust in that it can be applied to many different systems. For example, MRI has been used to visualize the pore space (Rigby and Gladden, 1996; Sederman et al.,

2001), to image nonaqueous phase liquids in porous media (Reeves and Chudek, 2001; Johns and Gladden, 1999), and to measure the flow velocity and the self diffusion coefficient of water with (Grenier et al., 1997; Irwin et al., 1999) and without (Baumann et al., 2000b; Ogawa et al., 2001) the use of paramagnetic tracers. MRI is also robust in that high resolution images can be obtained in short time periods. Typical imaging times with spin-echo sequences are in the range of less than one minute. By using echo-planar imaging sequences, the time for one single image can be reduced to fractions of a second (Vlaardingerbroek and den Boer, 1999). However, the latter technique is more sensitive to susceptibility artifacts, for example local blurring of the image caused by magnetic of paramagnetic components of the sediment matrix.

MRI was used to quantify colloid transport phenomena in porous media (Baumann and Werth, 2005). No such application was available at that time. The study illustrates how MRI can be used to resolve colloid transport mechanisms *in situ,* and to understand how transport at the column scale is affected by transport at finer scales.

As an example, series of images at different times during breakthrough are shown in Fig. 3.5 for a column filled with silica gel. Flow is from the left to the right. The color scale from blue to red indicates higher concentrations of the colloids. With increasing time, the effect of dispersion and flow variabilities can be seen as the colloids become more spread out. Also some colloids are retained at the column entrance near the Teflon® frit. It is not clear, whether this retention is due to the frit itself, or to variations of the silica gel packing density near the frit. With increasing time, the effect of the flow variability can also be seen. The colloids in the lower part of the column are moving faster than in the upper part of the column. All these effects would not be evident from conventional BTC data.

From the MRI images space-localized BTCs can be derived with highresolution (Fig. 3.6). In comparison to conventional breakthrough curves, this is a huge step forward in the course of providing data for process understanding. As an example, Fig. 3.6 summarizes the concentration profiles for superparamagnetic colloids in silica-gel at different positions inside the column together with the model fits using a numerical model (Toride et al., 1999). This example shows the power of MRI as an analytical tool to improve our understanding of colloid transport processes.

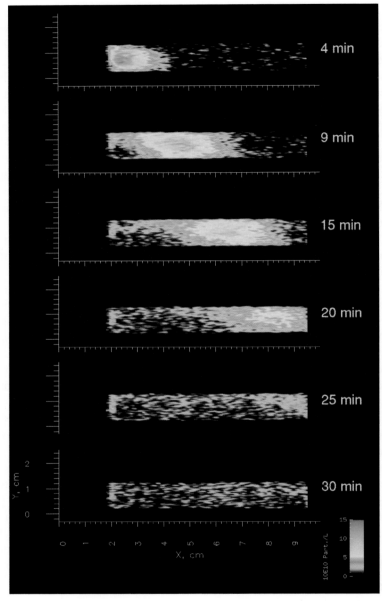

Fig. 3.5. Propagation of superparamagnetic colloids through column, flow rate = 9 mL/h, $C_0 = 1.3 \cdot 10^{12}$/L (Baumann and Werth 2005, Copyright with permission from Elsevier).

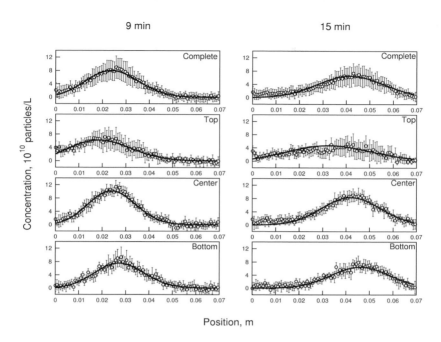

Fig. 3.6. Concentration profiles for selected parts of the central slice of a homogeneous column at different times. The error bars give the standard deviation of the colloid concentration perpendicular to the flow direction, the thick lines show the result of the CXTFIT approximation (Baumann and Werth 2005, Copyright with permission from Elsevier).

3.3.2 Assessment of Colloid Mass Transfer in Laboratory Columns

Previously described column tests suggested that colloids may act as carriers for heavy metal ions. In contrast to studies at the Gorleben site (Artinger et al., 1998; Kim et al., 1992), where the colloids were carefully equilibrated with the aquifer material, this set of column experiments (Ivanovich et al., 1996; Klein et al., 1996) was dedicated to the simultaneous input of colloids and contaminants. Neither the colloids nor the contaminants were in equilibrium with the aquifer material.

Fig. 3.7 shows the breakthrough of silica colloids through a column filled with quarternary gravel. The input concentration was kept constant at $8 \cdot 10^{14}$ particles/L. The breakthrough starts after 10 pore volumes and reaches a maximum of $1 \cdot 10^{13}$, or 1.25% of the input concentration. The overall mass transfer during the experiment was 38% of the input mass indicating a high filtration efficiency of the sediment. Silica colloids were found in aggregates after passing the column (Baumann et al., 1998).

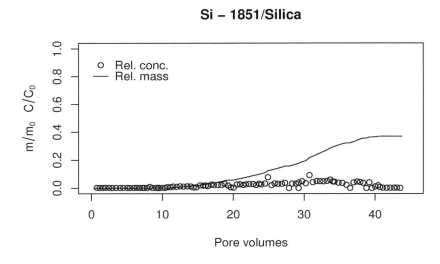

Fig. 3.7. Breakthrough curve for silica colloids through a quarternary sediment

Humic colloids (input conc. 2.5 mg/L) showed a two step pattern. First, they were filtered less effectively with effluent concentrations up to 32% of the input concentration. Afterwards, the concentrations decreased and the overall recovery rate of the colloids was about 20% of the input mass. The humic colloids were found to build up coatings on the calcareous sand and gravel within the first centimeters of the column. This indicates filter ripening. In both cases, the maximum velocity of the colloids was slightly faster compared to the mean velocity of a conservative tracer. This shows that pore scale processes like charge exclusion and size exclusion are influencing the transport of colloids. However, this example also proves, that the mass transfer rates of colloids have to be taken into account for the assessment of contaminant transport.

A set of column tests with sediments from the *Munich Gravel* Plain (calcareous gravel) and the *Sengenthal Fly Sand* Basin (quartz sand) was used to gain insight into the mass transfer rates of heavy metals associated with natural colloids. The input scenario was like described before: a spill of dissolved contaminants equilibrating with groundwater colloids.

The sediment columns were flushed with As(V), Cr(VI), and Ni(II) at constant concentrations of 1 mg/L and 10 mg/L. The effluent was filtered using ultrafiltration and the heavy metal concentration was determined for each colloid size class.

The colloid-associated heavy metals were usually found in the ng/mg concentration range. The concentration at the colloids scaled linear with the input concentration, except for As(V) and Cr(VI) in the Munich gravel. The distribution of the heavy metals to the colloids was mono-modal with a maximum in the 0.1-1 μm range (Fig. 3.8). The concentration at the Sengenthal colloids was higher compared to the Munich colloids for the lower input concentration (1 mg/L). This is attributed to a higher fraction of clay particles. At an in input concentration of 10 mg/L the concentration at the Sengenthal colloids was lower compared to the Munich colloids. This is attributed to a Langmuir-type sorption isotherm and limited sorption sites.

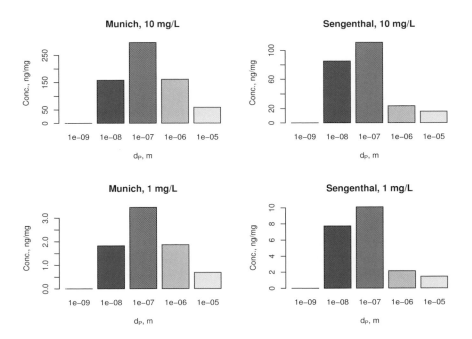

Fig. 3.8. Colloid associated As(V) concentrations

These two sets of experiments show the prerequisites of colloidal transport clearly: the contaminant has to establish a strong binding to the colloid, and the colloid itself has to be mobile. For the example of the Munich Gravel Plain, the concentrations of colloid associated heavy metal ions are in the upper ng/L range for natural colloid concentrations about 10 mg/L. This is well below the equilibrium concentrations of these heavy metal ions in a colloid free system and also well below the threshold values of the drinking water directive. If the colloid mobility and filtration effects are taken into account, the mass transfer rate is decreasing further.

3.4 Pilot Scale Experiments

3.4.1 Colloidal Transport on the Pilot Scale

A set of pilot scale aquifer tank experiments were performed to gain insight into the mass transfer of several heavy metal ions, and pentachlorophenol (PCP) under well controlled conditions. The sediments were taken from a heterogeneous gravel aquifer and a homogeneous sandy aquifer. The experiments simulate a spill of dissolved contaminants, and their subsequent transport with in-situ mobilized colloids that are not yet in equilibrium with the contaminant (Baumann et al., 2002a).

The colloid composition in the tanks was influenced by the tap water, which was slightly over-saturated with carbonate (saturation index for $CaCO_3 = 0.4$). Pressure changes at the interface to the mixing chamber of the tanks caused a degassing of carbon dioxide, followed by the precipitation of small, almost spherical calcite colloids (diameter < 100 nm). The calcite colloids were observed up to a flow distance of 2m. A natural equilibrium was established within the first 4 meters (Fig. 3.9). The chemical composition was dominated by carbonate colloids. Minor concentrations of silicates (mainly clay minerals), often as aggregates, were also detected. Colloids with diameters > 1 μm are slightly more frequent. The overall colloid concentration was approximately 3.5 mg L^{-1}.

In the *Sengenthal Sand* the calcite colloids were filtered out within the first 2 m (Fig. 3.9). The concentration of larger particles from $10 - 100$ μm increased with increasing flow distance. Colloids with a diameter < 10 μm were filtered out while passing through the tank. The total colloid mass concentration decreased from 6.5 mg L^{-1} to 2.5 mg L^{-1}. The colloids consisted predominantly of silicates (quartz, mica, feldspar).

The total mass of colloids present in both aquifer tanks is comparable to natural conditions at the corresponding field sites. The concentrations are, however, in the lower concentration range and cannot be compared to the colloid population in the vicinity of contaminated sites.

Fig. 3.10 shows the BTC of Ni(II) in *Munich Gravel* at a flow distance of 0.5 m. The transport occured predominantly in dissolved form, with the fraction associated with colloids at only 12%. The maximum concentration of the colloid-associated Ni(II) (at 1.9 h) was observed well before the maximum concentration of dissolved Ni(II) (3.5 h). However, the concentration of dissolved Ni(II) at $t = 1.9$ h was three times the concentration of colloidal transported Ni(II).

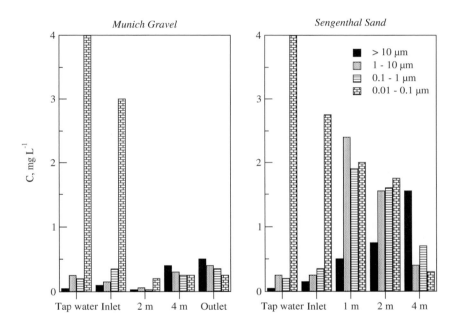

Fig. 3.9. Colloid size distribution (*Munich Gravel* and *Sengenthal Sand*) (Baumann et al. 2002a, Copyright with permission from Elsevier).

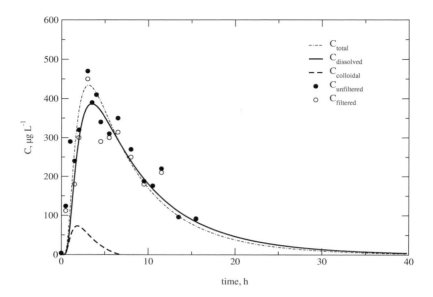

Fig. 3.10. Breakthrough curve of Ni(II) at a flow distance of 0.5 m, *Munich Gravel* (Baumann et al. 2002a, Copyright with permission from Elsevier).

PCP in *Sengenthal Sand* at a flow distance of 2 m showed a sharp BTC with almost no tailing (Fig. 3.11). The recovery was 23% and the fraction associated with colloids was 22%. There was no significant difference between the transport velocity of the contaminant and that of the conservative tracer.

The fraction associated with colloids in general increases moderately with increasing flow distance. While those contaminants, which are characterized by moderate sorption to the sediment, retain their distribution coefficient f_{coll}, the contaminants like Cu^{2+}, which are strongly sorbing to the matrix show an increasing f_{coll}. Contaminants associated with colloidal material are no longer subject to fluid–surface interactions, their transport is controlled by the transport of the colloidal material itself. If the fluid–surface interaction is moderate, as for PCP, and if the sorption and desorption kinetics are sufficiently fast, a dynamic equilibrium between fluid, sediment surface, and colloid is established. The equilibrium then depends on the actual concentrations of the contaminant within the three-phase system. Therefore, the fraction associated with colloids will become independent of the flow distance.

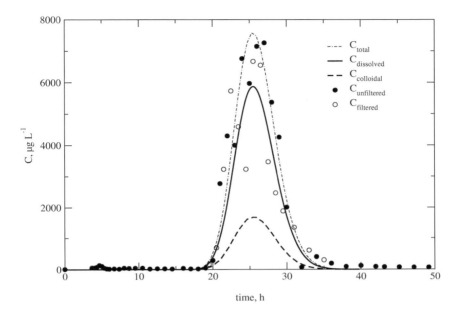

Fig. 3.11. Breakthrough curve of PCP at a flow distance of 2 m, *Sengenthal Sand* (Baumann et al. 2002a, Copyright with permission from Elsevier).

The effective filtration in *Munich Gravel* restricts the mass transfer of contaminants associated with colloids because the colloids themselves are not very mobile.

With regard to the experimental setup, the results show the pitfalls one may experience during the interpretation of breakthrough curves with low recovery rates. In principle low recovery rates alone do not raise difficulties during the evaluation, as long as the signal-to-noise ratio (S/N) is still adequate. The S/N ratio honors all errors from sampling, sample preparation, and analysis. Additional statistical significance can be gained from the repetition of the experiments (as within the presented study) and from additional sampling points in time and space. However, if the concentrations drop below the limits of detection the experimental design has to be modified (e.g. to use a step input function or higher input concentrations, if possible). In any case, it is important to note that the obtained values are representative of the mobile fractions of the contaminants or colloids. A direct comparison to the transport parameters of the water or a conservative tracer is not appropriate.

The results acquired during the pilot scale migration experiments are in agreement with the theoretical framework of colloidal transport described in the literature. With increasing K_D values and increasing flow distance the relative fraction of the contaminant transported with colloids increases. However, the colloids themselves are subject to matrix interactions, and therefore overall mass transfer is limited. If the sorption/desorption kinetics are significantly fast and if the K_D value is low, then a dynamic equilibrium between solution and colloids is established. In that case, the fraction of contaminants bound to colloids remains the same, regardless of the transport length. This behavior was observed for PCP. This finding implies that on a field scale a significant fraction of the contaminant can be found to be associated with colloids, even at locations which are far beyond the maximum flow distances for the colloids given by filtration theory. One critical issue that remains to be investigated is the accurate assessment of the heterogeneity of the aquifer at the field scale. The experiments with the *Munich Gravel* already illustrate that on a flow length of ten meters preferential flow paths heavily influence the transport of colloids and contaminants.

3.5 Colloid Transport on the Field Scale

3.5.1 Colloidal Transport in Uncontaminated Environments

Following the transport experiments on the column and pilot scale, it was of great interest to obtain data from field sites under natural conditions. Again, the input scenario was a spill of dissolved contaminants into groundwater (Huber et al., 1998; Baumann et al., 2000a).

The breakthrough of As(V) after a flow distance of 38m is compared to the breakthrough of uranine in Fig. 3.12. As(V) is transported significantly slower compared to the Tracer. The total recovery of As(V) was 1.4% of the input mass, roughly 17% of the mobile As(V) was associated to colloids.

A completely different picture was obtained for Cr(VI) (Fig. 3.13, which followed the tracer breakthrough. The recovery rate was 81%, the fraction associated to colloids was 6%. These findings were expected due to the formation of hydrato-complexes of Cr(VI). Field experiments in igneous rocks at the Nabburg site showed similar results (Baumann et al., 2000a).

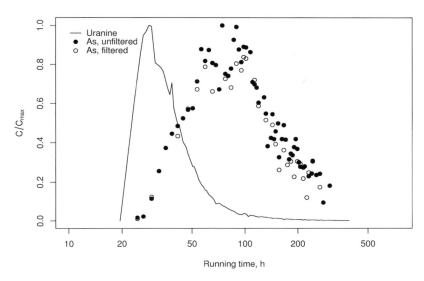

Fig. 3.12. BTC for As(V) and uranine in a quarternary gravel

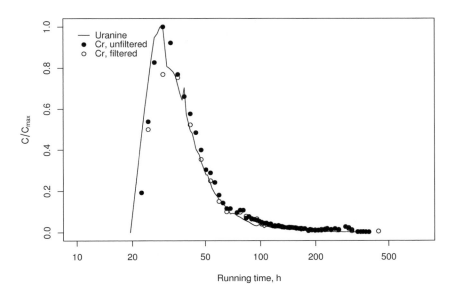

Fig. 3.13. BTC for Cr(VI) and uranine in a quarternary gravel

The results of the field experiments indicate that a dynamic equilibrium between solution, colloids, and matrix has to be assumed under natural conditions. Within a contamination plume, the contaminant concentration associated to colloids is a function of the concentration of the contaminant in the solution, since the transport distance of the colloids is limited by filtration effects. The fraction of colloid associated contaminants increases with increasing transport distance if the equilibrium of the contaminant is shifted to the solid phase and the colloids are mobile. However, at typical colloid concentrations in shallow aquifers, the colloidal mass transfer is limited.

The field experiments also illustrate that upscaling from pilot experiments to field experiments is possible. However, as long as the processes at the pore scale are not fully understood, therefore necessitating fitting parameters to lump the heterogeneity of the setup and the sediment, the results might deviate significantly from the predictions.

3.5.2 Colloidal Transport in Contaminated Environments

There are still many landfills without proper sealing systems and emitting toxic substances such as heavy metals into groundwater. Heavy metals are well known to bind to colloidal matter present in leachate and groundwater (Gounaris et al., 1993; Baun and Christensen, 2004). Heavy metals associated with colloids may show significantly different transport properties compared to dissolved heavy metals. Here, the transport of heavy metals in the presence of colloids may be enhanced or retarded (McCarthy and Zachara, 1989; Kretzschmar et al., 1999), depending on the hydrogeological and hydrochemical conditions. Thus, the presence of colloids in landfill leachate contaminating the groundwater may alter the propagation of heavy metals in the contamination plume.

Colloids and colloid transport are very sensitive to hydrochemical and hydrodynamic conditions (Roy and Dzombak, 1997; Bergendahl and Grasso, 2000; Bradford et al., 2002; Ryan and Elimelech, 1996). These changes are likely to occur in the vicinity of landfills (Bjerg et al., 1995). Thus, both mobilisation of natural colloids and precipitation of colloids from landfill leachate are likely downgradient of landfills.

In leachate from four Danish landfills generally more than 60% of the total heavy metal concentrations were associated to colloids (Jensen and Christensen, 1999). In contaminated groundwater close to landfills, spiked heavy metal ions were found to bind to colloids (Jensen et al., 1999). From these results, a transfer of colloid associated heavy metals from the landfill into the groundwater does not seem unlikely. Despite the potential effects on the assessment of heavy metal transport downgradient of landfills there are only a few field studies.

A recent review (Baun and Christensen, 2004) listed only 8 field studies addressing the speciation of heavy metals in landfill leachate, five of them also addressing the distribution of heavy metals between colloids and dissolved fractions. In general, organic colloids were found to be important with respect to heavy metals. A possible importance was stated for precipitates (iron, sulphide, carbonate, and phosphate), the importance of natural colloidal matter (clay, quartz) was not clear.

One issue which remains open in the current literature is the transport from the landfill into the groundwater. From the current state of knowledge, it seems likely that colloids and heavy metals are emitted into the groundwater together. It might also be possible that colloid bound heavy metals are filtered out at the interface between landfill and groundwater, thus leading to a retardation of the colloids and the heavy metals. Finally, dissolved heavy metals emitted from the landfill might bind to colloids in the groundwater.

The fact that a large fraction of the heavy metals in landfill leachate as well as in groundwater are bound to colloids does not necessarily prove that there is colloidal transport between landfill and groundwater, although this seems to be straightforward.

During several research projects (Klein et al., 1996; Ivanovich et al., 1996; Huber et al., 1998), the role of colloids in the vicinity of landfills with direct contact and in shallow groundwater has been addressed.

In the vicinity of landfills, the comparison of the particle size distribution between leachate and downgradient groundwater shows significant differences. This is a first indication that the release of colloids into the groundwater is incomplete (Baumann et al., 2006).

The mineralogical characterization of the colloids in the vicinity of the Augsburg disposal is documented in Fig. 3.14. The size class from 10nm to 100nm consists of salt colloids, iron oxides and iron hydroxides and a significant fraction of not classified colloids (e.g. clay minerals). The high fraction of not classified colloids is an intrinsic limitation of the SEM/EDX-technique applied to small particle sizes. Consequently, the fractions of not classified colloids decrease with increasing particle size. Carbonaceous colloids and silicates are dominating the larger colloid size classes. In summary, upgradient of the sites the colloids mirror the geochemical composition of the aquifer matrix.

Downgradient of the disposal the smallest size class (10 nm to 100 nm) is again dominated by salt colloids and not classified or multiple classified colloids. Iron colloids were determined at the Augsburg disposal. The colloid size class from 100 nm to 1 μm contains almost all mineralogical classes. Salt colloids and iron oxides and iron hydroxides are dominating at the Augsburg disposal, iron colloids are prevailing in the larger colloid size classes. The high concentration of iron colloids is due to the redox potential changes downgradient of the disposal, causing a precipitation of iron colloids.

Inside of the Augsburg disposal the colloids in the smallest size class are dominated by organic colloids and salt colloids. Carbonates, silicates and arsenic containing colloids are prevailing in the larger size classes.

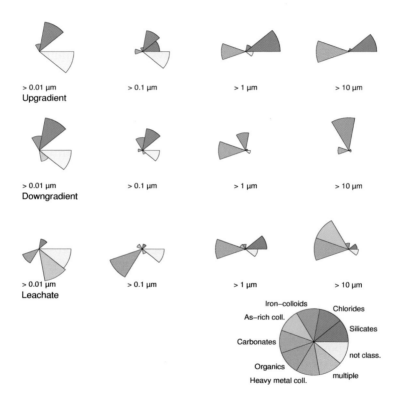

Fig. 3.14. Chemical characterization of the colloids at the Augsburg disposal (Baumann et al. 2006, Copyright with permission from Elsevier).

In principle, the colloids upgradient and downgradient of the disposals are comparable. However, the colloids inside of the disposals are significantly different. Together with a surprisingly low concentration of organic colloids and an insignificant fraction of heavy metal containing colloids downgradient of the disposals, this finding points to an insignificant transfer of colloidal matter from the disposal to the groundwater. This is in contrast to other studies where a high concentration of organic colloids were found (Baun and Christensen, 2004).

There are two reasons for this observation: First, the SEM/EDX technique is not sensitive for the smallest colloids and this is the size range where fulvic and humic substances are predominating. However, looking at the size distribution, these size classes do not contribute much to the groundwater colloids at all sites and the leachate at the Gallenbach disposal. Second, organic colloids tend to interact with the carbonaceous matrix of the aquifer. Evidence for this assumption was found in boreholes close to the Augsburg disposal. The sediment matrix was found to be covered completely with organic coatings. Additionally, redox potential changes cause a precipitation of iron, manganese and arsenic when the leachate propagates from the reducing conditions inside of the disposals to the oxidizing conditions downgradient of the disposals. The formation of colloids, in this case, depends on the extent of the contamination plume.

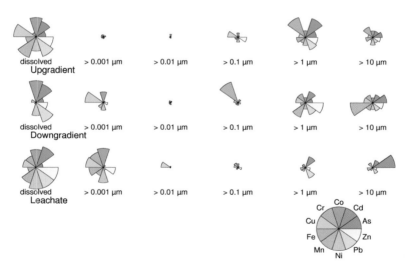

Fig. 3.15. Association of heavy metals to colloids at the Augsburg disposal (Baumann et al. 2006, Copyright with permission from Elsevier).

The distribution pattern of the heavy metal-colloid association reveals significant differences between the leachate in the disposal and the downgradient groundwater (Fig. 3.15). Again, there is evidence for a precipitation of iron and arsenic due to hydrochemical changes in the contamination plume.

The results indicate that there is no direct transfer of colloids from the disposal sites into the aquifers. Field investigations suggest that colloids are filtered at the interface between waste disposal. The deposited colloids are forming coatings which might increase the sorption capacity of the sediment and decrease the hydraulic conductivity of the interface. The mobile colloids in the size classes between 10 nm and 1 μm were found to bear only very low heavy metal concentrations. Thus, there is no proof for a colloid enhanced heavy metal transport at the investigated sites. The change of the hydrochemical conditions at the interface between landfill and groundwater seems to be an effective barrier for colloids at these sites. Heavy metals attached to colloids are likely to get immobilized together with the colloids. However, due to the changes of the redox potential conditions colloids might be generated from supersaturated solutions. These colloids might then propagate through the aquifer. The formation of organic and inorganic coatings at the interface leads to reduced hydraulic conductivity values and increased sorption efficiencies. These changes reduce the overall emissions from the landfill.

3.6 Outlook

The transport processes of colloids are seen clearer now. However, a complete physical, chemical and microbiological framework for the prediction of colloidal transport phenomena is still missing. This is partly due to the complexity of natural environments.

Some potential colloid transport scenarios have received only limited attention so far: colloid transport in multiphase systems, eg. the vadose zone of aquifers, are currently under investigation. Here, micromodel experiments and MRI provide the analytical techniques to assess the dynamics of such systems. Assessment of colloid-mediated transport at heavily contaminated sites requires the consideration of transient collectors, eg. nonaqueous liquids, which is not yet the case. The effects of biofilms on colloid transport have largely been suppressed, by maintaining sterile conditions in the experiments. It is obvious, that an aquifer is far from being sterile. Cooperation with microbiologists and the use of methods to detect the formation of organic coatings on the substrate will bring light into these phenomena.

The transport of colloids is not only affected by the properties of the matrix. It seems very likely that the transport of bacteria in porous media is significantly different, if the bacteria are emitted with manure, compared to an injection to pure bacteria cultures in a column test. Field-flow-fractionation techniques with reactive membranes might prove a valuable tool to assess the transport properties of colloids under natural conditions.

Another upcoming question is the fate of colloids at contaminated sites during and after site remediation. Our theoretical knowledge suggests a colloid mobilization due to changes of the hydraulic and hydrochemical conditions during site remediation. In situ evidence for this assumption is still missing because at the few sites where colloid transport has been studied extensively, the hydrochemical conditions have not yet returned to normal.

Finally, the majority of colloid transport studies have been conducted under humid conditions. It is questionable whether the results can be transferred to arid environments. Available studies suggest, that in situ colloid formation and immobilisation due to weathering effects is much more pronounced under arid and semi-arid conditions. It will be interesting to run experiments like this under MRI or synchrotron inspection.

In the meantime, despite possible scientific shortcomings, methods for a quick assessment of colloid transport phenomena for everyday use have to be developed. A rule based fuzzy model to predict the probability of colloid enhanced transport based on the hydrogeological and hydrochemical settings seems to be a feasible tool.

In conclusion, the research on colloid mediated transport has not yet reached an end. Future research will be dedicated to a better process understanding using advanced analytical and experimental techniques. On the other hand, there is still need for further pilot scale and field scale studies to validate the theoretical assumptions. Ostwald's "world of neglected dimensions" will still receive significant scientific attention.

References

Artinger R, Kienzler B, Schussler W, Kim JI (1998) Effects of humic substances on the am-241 migration in a sandy aquifer: column experiments with Gorleben groundwater/sediment systems. J Contam Hydrol 35:261–275

Backhus DA, Ryan JN, Groher DM, MacFarlane JK, Gschwend PM (1993) Sampling colloids and colloid-associated contaminants in ground water. Ground Water 31:466–479

Baumann T, Fruhstorfer P, Klein T, Niessner R (2006) Colloid and heavy metal transport at landfill sites in direct contact with groundwater. Water Research 40:2776-2786

Baumann T, Fruhstorfer P, Niessner R (1998) Sickerwassertransport von Kolloiden und Schwermetallen in Bergbaufolgelandschaften – Felduntersuchungen und Laborversuche. Grundwasser 3:3–13

Baumann T, Huber N, Müller S, Niessner R (2000a) Quantification of the colloidal transport through laboratory and field experiments. Vom Wasser 95:151–166

Baumann T, Müller S, Niessner R (2002a) Migration of dissolved heavy metal compounds and PCP in the presence of colloids through a heterogeneous calcareous gravel and a homogeneous quartz sand – Pilot scale experiments. Water Research 36:1213–1223

Baumann T, Niessner R (2006) Micromodel study on repartitioning phenomena of a strongly hydrophobic fluorophore at a colloid/1-octanol interface. Water Resour. Res. 42:W12S04

Baumann T, Petsch R, Niessner R (2000b) Direct 3-D measurement of the flow-velocity in porous media using magnetic resonance tomography. Environ Sci Technol 34:4242–4248

Baumann T, Werth CJ (2004) Visualisation and modelling of polystyrol colloid transport in a silicon micromodel. Vadose Zone Journal 3:434–443

Baumann T, Werth CJ (2005) Visualization of colloid transport through heterogeneous porous media using magnetic resonance imaging. Colloids and Surfaces, A: Physicochemical and Engineering Aspects 265:2–10

Baumann T, Werth CJ, Niessner R (2002b) Visualisation of colloid transport processes with magnetic resonance imaging and in etched silicon micromodels. DIAS Report Plant Prod 80:25–30

Baun DL, Christensen TH (2004) Speciation of heavy metals in landfill leachate: a review. Waste Managem Res 22:3–23

Bergendahl J, Grasso D (2000) Prediction of colloid detachment in a model porous media: Hydrodynamics. Chem Eng Sci 55:1523–1532

Bjerg PL, Rügge K, Pedersen JK, Christensen TH (1995) Distribution of redox-sensitive groundwater quality parameters downgradient of a landfill (Grindsted, Denmark). Environ Sci Technol 29:1387–1394

Bradford SA, Yates SR, Bettahar M, Simunek J (2002) Physical factors affecting the transport and fate of colloids in saturated porous media. Water Resour Res 38:1327–1338

Crist JT, McCarthy JF, Zevi Y, Baveye P, Throop JA, Steenhuis TS (2004) Pore-scale visualization of colloid transport and retention in partly saturated porous media. Vadose Zone Journal 3:444–450

Crist JT, Zevi Y, McCarthy JF, Throop JA, Steenhuis TS (2005) Transport and retention mechanisms of colloids in partially saturated porous media. Vadose Zone Journal 4:184–195

Gao B, Saiers JE, Ryan JN (2004) Deposition and mobilization of clay colloids in unsaturated porous media. Water Resour Res 40:W08602

Gounaris V, Anderson PR, Holsen TM (1993) Characteristics and environmental significance of colloids in landfill leachate. Environ Sci Technol 27:1381–1387

Grenier A, Schreiber W, Brix G, Kinzelbach W (1997) Magnetic resonance imaging of paramagnetic tracers in porous media: Quantification of flow and transport parameters. Water Resour Res 33:1461–1473

Greswell R, Lloyd JW, Tellam JH, Parker D (1998) The study of solute movement through rock using positron emission tomography. Ann Geophys 16, C 439

Happel J (1958) Viscous flow in multiparticle systems: Slow motion of fluids relative to beds of spherical particles. AIChE J 4:197–201

Huber N, Müller S, Baumann T, Niessner R (1998) Ausbreitung deponierelevanter Substanzen in unterschiedlichen Grundwasserleitern. Report E 18, Bavarian Water Management Agency, D–80636 München

Irwin NC, Altobelli SA, Greenkorn RA (1999) Concentration and velocity field measurements by magnetic resonance imaging in aperiodic heterogeneous porous media. Magn Reson Imaging 17:909–917

Ivanovich M, Baumann T, Tellam JH, Read D (1996) The role of colloids in the transport of pollutants in groundwaters: Development of monitoring and assessment procedures. AEA Technology, Harwell

Jensen DL, Christensen TH (1999) Colloidal and dissolved metals in leachates from four danish landfills. Water Research 33:2139–2147

Jensen DL, Ledin A, Christensen TH (1999) Speciation of heavy metals in landfill-leachate polluted groundwater. Water Research 33:2642–2650

Jia C, Shing K, Yortsos YC (1999) Visualization and simulation of non-aqueous phase liquids solubilization in pore networks. J Contam Hydrol 35:363–387

Johns ML, Gladden LF (1999) Magnetic resonance imaging study of the dissolution kinetics of octanol in porous media. J Colloid Interface Sci 210:261–270

Kim J, Zeh P, Delakowitz B (1992) Chemical interactions of actinide ions with groundwater colloids in Gorleben aquifer systems. Radiochim Acta 58/59:147–154

Klein T, Baumann T, Fruhstorfer P, Fies M, Niessner R (1996) Untersuchung der Kolloide als Träger für den Transport von Schadstoffen in Sicker- und Grundwässern. Schlußbericht BayFORREST F20, Bavarian State Ministry for Regional Development and Environmental Affairs, 81925 München

Knutson CE, Werth CJ, Valocchi AJ (2001) Pore-scale modeling of dissolution from variably distributed nonaqueous phase liquid blobs. Water Resourc Res 35:2951–2963

Kretzschmar R, Borkovec M, Grolimund D, Elimelech M (1999) Mobile subsurface colloids and their role in contaminant transport. Adv Agron 66:121–193

McCarthy JF, Zachara JM (1989) Subsurface transport of contaminants. Environ Sci Technol 23:496–502

Megens M, van Kats CM, Böseke P, Vos WL (1997) In situ characterization of colloidal spheres by synchrotron small-angle x-ray scattering. Langmuir 13:6120–6129

Neuhäusler U, Abend S, Jacobsen C, Lagaly G (1999) Soft x-ray spectromicroscopy on solid-stabilized emulsions. Colloid Polym Sci 277:719–726

Ogawa K, Matsuka T, Hirai S, Okazaki K (2001) Three-dimensional velocity measurement of complex interstitial flows through water-saturated porous media by the tagging method in the MRI technique. Meas Sci Technol 12:172–180

Rajagopalan R, Chu RQ (1982) Dynamics of adsorption of colloidal particles in packed beds. J Colloid Interface Sci 86:299–317

Rajagopalan R, Tien C (1976) Trajectory analysis of deep bed filtration with the sphere-in-cell porous media model. AIChE J 22:523–533

Reeves AD, Chudek JA (2001) Nuclear magnetic resonance imaging (MRI) of diesel oil migration in estuarine sediment samples. J Ind Microbiol Biotechnol 26:77–82

Rigby SP, Gladden LF (1996) Nmr and fractal modelling studies of transport in porous media. Chem Eng Sci 51:2263–2272

Roy SB, Dzombak DA (1997) Chemical factors influencing colloidfacilitated transport of contaminants in porous media. Environ Sci Technol 31:656–664

Ryan JN, Elimelech M (1996) Colloid mobilization and transport in groundwater. Colloids and Surfaces, A: Physicochemical and Engineering Aspects 107:1–56

Sahloul NA, Ioannidis MA, Chatzis I (2002) Dissolution of residual non-aqueous phase liquids in porous media: pore-scale mechanisms and mass transfer rates. Adv Water Resour 25:33–49

Saiers JE, Hornberger GM, Gower DB, Herman JS (2003) The role of moving air-water interfaces in colloid mobilization within the vadose zone. Geophys Res Lett 30:2083

Saiers JE, Lenhart JJ (2003a) Colloid mobilization and transport within unsaturated porous media under transient-flow conditions. Water Resour Res 39:1019

Saiers JE, Lenhart JJ (2003b) Ionic-strength effects on colloid transport and interfacial reactions in partially saturated porous media. Water Resour Res 39:1256

Sederman AJ, Alexander P, Gladden LF (2001) Structure of packed beds probed by magnetic resonance imaging. Powder Technol 117:255– 269

Soll WE, Celia MA, Wilson JL (1993) Micromodel studies of 3-fluid porous-media systems - pore-scale processes relating to capillarypressure saturation relationships. Water Resour Res 29:2963–2974

Sugita F, Gillham RW (1995) Pore scale variation in retardation as a cause of non-ideal reactive breakthrough curves, 3, column investigations. Water Resour Res 31:121–128

Toride N, Leij FJ, van Genuchten MT (1999) The CXTFIT code for estimating transport parameters from laboratory or field tracer experiments, version 2.1. Research Report 137, U.S. Salinity Lab., Agric Res Serv, U.S. Dep. of Agric., Riverside, Calif

Tufenkji N, Elimelech M (2004) Correlation equation for predicting single-collector efficiency in physicochemical filtration in saturated porous media. Environ Sci Technol 38:529–536

Vlaardingerbroek MT, den Boer JA (1999) Magnetic resonance imaging. Springer, Berlin

Wan J, Wilson JL (1994a) Colloid transport in unsaturated porous media. Water Resour Res 30:857–864

Wan J, Wilson JL (1994b) Visualization of the role of the gas-water interface on the fate and transport of colloids in porous media. Water Resour Res 30:11–23

Wan JM, Tokunaga TK, Tsang CF, Bodvarsson GS (1996) Improved glass micromodel methods for studies of flow and transport in fractured porous media. Water Resour Res 32:1955–1964

Weisbrod N, Ronen D, Nativ R (1996) New method for sampling groundwater colloids under natural gradient flow conditions. Environ Sci Technol 30:3094–3101

4 Transport of Colloids in Filter Columns: Laboratory and Field Experiments

Ralf Siepmann[1], Frank von der Kammer[2], Ulrich Förstner[1]

[1] Institute of Environmental Technology and Energy Economics, Hamburg Univeristy of Technology, Germany
[2] Center for Earth Sciences, University of Vienna, Austria

4.1 Introduction

Urbanization, the development and improvement of infrastructure leads to a steady increase of the fraction of sealed surfaces. The direct consequence is an increasing volume of surface runoff producing relevant loads of contaminants. The problem arising must be addressed in terms of runoff volume, rate characteristics (low retention) as well as water quality. Currently, most of the runoff is infiltrated, collected in dual urban sewer systems and discharged without further treatment or is partly passed through water treatment plants where it is cleaned up with comparably expensive treatment methods. Direct seepage into soils or infiltration ditches is commonly favoured as an alternative, since it supports the desired groundwater recharge. However, persistent contaminants that are ubiquitous in these runoff waters will not be permanently stored in filters or soils, but may migrate slowly towards groundwater, causing a long-term potential threat for the degradation of groundwater quality (Pitt et al. 1996). The present study deals with the transport potential of solids in runoff with respect to the transfer of micropollutants into the groundwater zone as well as with the efficiency of advanced filtration facilities to remove small-sized pollutant-loaded particles.

Atmospheric fine dust is considered as a main source of the runoff pollution. It generally consists of particles in or near the colloidal size fraction (PM 2,5 and PN 0,5), that carry the highest concentrations of heavy metals (Allen et al. 2001; Harrison et al. 2003; Pakkanen et al. 2003). For example, in mixed waters single particles and aggregates have been found composed of sulphur compounds originating from brake wear, rare earth metals from catalyst exhaust, SnPb-alloys from roof plating and lead carbonates from wall paintings (Samrani et al. 2004). Tire and brake wear and products of the car body corrosion seem to be equally important sources for contamination of the runoff (Sternbeck et al. 2002).

A rough exemplary calculation of the annual amount of solids and heavy metals deposited along the highly frequented part of the roads in Germany shows that the problem of diffuse pollution is much worse than one would expect: calculations based on data from the Federal Statistical Office of Germany for the area consumed by main roads and highways in 2001 and the annual precipitation of that year, combined with the detected solids and metal contents of runoff samples (Kasting 2001, Meißner 2003), resulted in the data shown in Table 4.1.

Table 4.1. Calculation of annual amounts of solids and metals on German federal roads using road area and precipitation

content:	mean runoff	total annual	per road-km
solids total	75.0 mg·L^{-1}	460991 t	1.9 t
Zn	0.46 mg·L^{-1}	2827 t	11.6 Kg
Cu	0.11 mg·L^{-1}	676 t	2.8 Kg
Pb	0.04 mg·L^{-1}	246 t	1.0 Kg

A newer study of the the Federal Environment Agency of Germany (Hillenbrand et al. 2005) that is based on an emission calculation per vehicle differentiates the total annual contributions from the following sources:

- Tire wear: 111,420 t, containing 1,619 t Zn and 2.69 t Pb;
- Brake wear: 12,350 t, containing 380 t Zn, 927 t Cu, and 61.5 t Pb;
- Road wear: 1,737,120 t, containing 150 t Zn.

Accounting for other sources as well as wet and dry deposition both approaches obtain comparable results. The differences in "total solids" from Table 4.1 and "road wear" indicate that a significant portion of the particulate emissions from traffic are not found in road runoff.

A substantial proportion of this contamination will have to be removed from the runoff for fulfilling the up-coming national and international objectives for the preservation of soil and groundwater quality.

State-of-the art filtration consists of flat bed sand filters with or without plantation, vertical sand filters or different systems of direct infiltration into the vadose zone. Recent approaches for road runoff treatment are based on low-cost reactive filter materials such as natural zeolite, which removes dissolved heavy metals species by ion exchange (Jacobs et al. 2000; v.d. Kammer and Jacobs 2003).

The transport or retention of heavy metals in filtration columns used for the clean-up of urban runoff is largely influenced by the content and composition of colloidal matter already present in the runoff. The composition of runoff as well as the content of organic matter varies widely. Especially at rainstorm events, elevated concentrations of suspended particulate matter are recognized especially in the first-flush resulting in higher loads of particulate bound contaminants washed from the road which may also carried through retention ponds due to increased turbulence.

The heterogeneous mixture of colloidal particles present in road runoff offers a large amount of favourable sorption sites for cationic pollutants. Depending on seasonal influences, a variable fraction of heavy metals will be transported particle-bound, sometimes almost all. Direct seepage to groundwater level through adapted filtration plants must deal with the particles already present in the runoff, different hydraulic loads and resulting flow rates as well as a changing degree of water saturation.

Seepage via roadside soils is also influenced by the input of particles form the road. Additional source for colloid-bound transport is in-situ mobilization of particles from the soil matrix induced by hydraulic and hydrochemical gradients, such as varying load and water saturation, and ionic strength gradients induced mainly by the application of de-icing agents. The contaminants bound to such highly mobile colloids may be transported faster than those in solution (McCarthy and Zachara 1989, Förstner et al. 2001)

The main parameter for these gradients is the variability of ionic strength in road runoff under winter service, with sodium concentrations ranging from 2-16000 mg·L^{-1}. In the runoff samples which were collected at a federal road in Hamburg a relevant quantity of fine particles remains in stable suspension even at very high ionic strengths of several g·L^{-1} sodium chloride.

These facts make an effective filtration indispensable that retards both solved and particle-bound pollutants. To understand both the processes that favour the desired irreversible deposition of particles and colloids on a filter matrix and the ones that lead to remobilization or unaffected transport of particle-bound pollutants, we applied a two-track strategy:

1. On one side we studied the influence of the ionic strength in simplified laboratory-scale systems consisting of sand-filled columns and well-characterized colloidal suspensions under controlled conditions.
2. On the other side we analyzed the seasonal and event-induced runoff composition at a test site at a frequented road and ran a pilot-scale vertical sand filtration column fed with road runoff, to study the retention behaviour under complex field conditions.

Based on the findings from the laboratory (see Sect. 4.2) and field (see Sect. 4.3) studies, proposals for optimization of the filtration are made in Sect. 4.4.

4.2 Laboratory Studies of Colloidal Transport Behaviour

4.2.1 Used Materials

The colloids used for laboratory experiments consisted of long-term stabilized suspensions of natural and close-to-nature "reference colloids". They were obtained from extracting clay material and aquifer soils while conserving the natural properties of the materials (v.d. Kammer 2005; Baalousha et al. 2005).

The extraction sequence was developed and modified on the basis of existing procedures. It consists of the following steps, that can be adjusted according to the source material properties: (1) sieving, washing, pretreatment , (2) extraction of the colloidal fraction by shaking or ultrasonication. (3) centrifugation or sedimentation for the removal of coarse particle fractions, and (4) purification by centrifugation or dialysis.

The size distribution of the suspensions as characterized by field-flow fractionation (FFF) was in the range of 50 to 400 nm.

For the constant particle feed experiments, the morphologic composition of runoff samples was simulated with an extract of "illite clay", composed of 30% illite, 30% smectite, 25% quartz and 10 % kaolinite.

In addition to these clay-based heterogeneous mineral suspensions, a quartz suspension was obtained by extraction and centrifugation of ground quartz powder, with particles in a size range from 300 to 600 nm, determined by F^4 (see 4.2.2).

The artificial, industrially produced clay mineral laponite, a magnesium lithium silicate with the formula $[Na_{0.7}]^+ [(Si_8\ Mg_{5.5}\ Li_{0.3})\ O_{20}\ (OH)_4]^{0.7-}$, which consists of uniform platelets with a diameter of about 25 nm, was also used as a simplified clay mineral model colloid suspension. To match the sensitivity levels of the analytic instruments, the concentrations used in the laponite suspension were much higher ($g \cdot L^{-1}$) than in the natural samples ($10\ mg \cdot L^{-1}$).

4.2.2 Optimization and Adaption of Analytical Devices

a) Flow-Field Flow Fractionation (F^4) Method - Optimization and Particle Shape Analysis.

The application of F^4, an analytical fractionation method based on hydro-dynamic principles, for the separation and characterization of heterogeneous natural colloids and its advantages over batch-methods as e.g. photon correlation spectroscopy (PCS) is well documented. The principle of operation and its potentials as well as limitations are extensively described in the literature (for F^4: Beckett and Hart 1993; v.d. Kammer and Förstner 1998; v.d. Kammer 2005; v.d. Kammer et al. 2005a + b; Hassellov et al. 2005, for PCS: Schurtenberger and Newman 1993; Filella et al. 1997).

A critical step in method development for the F^4 analysis is the proper adaptation of carrier liquid to provide an unperturbed fractionation limiting effects of particle-particle and particle-wall interactions. For the analysis of heterogeneous natural colloids this adaptation should be performed always when a sample containing colloids of different composition and type is analysed. Since a thorough theory for the proper composition of carrier liquid (addition of dispersing agents, background electrolyte and ionic strength) is still lacking, it is essential to investigate the behaviour of the sample within the chosen carrier composition prior F^4 measurement and to apply means of independent particle size analysis on the fractions eluting from the F^4 channel to verify an unperturbed fractionation.

This has been achieved by:

- investigation of the influence of typical carrier compositions on the fractionation
- introduction of an optimization routine to select carrier composition for a maximum of negative zeta potential
- the proof that multi-angle laser light scattering detectors (MALLS) are suitable for the independent particle size determination in sample fractions eluting from F^4.

The latter enables a direct evaluation of the F^4 process.

Fig. 4.1 shows the influence of different concentrations of two detergents often used in FFF on the fractionation in terms of average particle size as the first moment of the size distribution and peak area as a proxy for the recovery in FFF. The sample used was a stable colloidal suspension of aquifer matrix colloids obtained as described before. For the detergent sodium dodecyl sulfate (SDS) in SedFFF (Sedimentation Field-Flow Fractionation) and F^4 there is a certain concentration for a maximum peak area (at about 0.025 %) and the average particle size also has its maximum value at this concentration (more larger particles eluting, but not necessarily an indication of aggregation). The Fl-70, a commercially available mixture of various cationic and non-ionic surfactants, seems to provide best performance at ≥ 0.05 % (peak area) in both F^4 and SedFFF. However in using Fl-70 the average particle size is not consistent between F^4 and SedFFF and increased concentrations of FL-70 seem to initiate some aggregation and/or wall interactions (SedFFF).

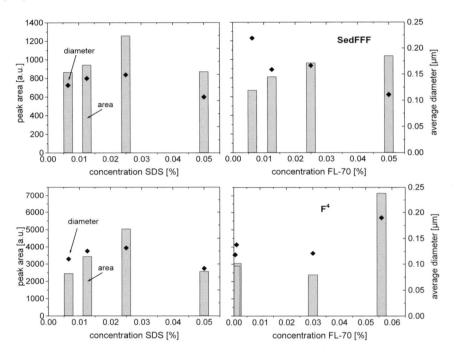

Fig. 4.1. Comparison of peak data retrieved for sample reference1 from both F^4 and SedFFF applying different concentrations of the surfactants SDS and FL-70. Since the larger particles are most influenced by particle-wall effects an increase of peak area may come along with an increase of average particle size (from v.d. Kammer (2005))

To find optimal carrier composition for SDS and FL-70 as surfactants and to investigate the influence of the surfactants on the stability of natural samples, zeta potential measurements should be performed always and were essential in the optimization process before F^4 analysis of unknown samples. As an example, the zeta potentials of a reference sample (same as in Fig. 4.1) were determined as a function of surfactant concentration (Fig. 4.2).

While FL-70 provides a distinct minimum at concentrations which are similar to those where largest peak area was obtained in FFF, SDS has only a little influence on zeta potential leading to aggregation of the sample above 0.2 %. The little decrease of zeta potential when concentration rises from 0 % to 0.1 % can explain the fractionation behavior which has an optimum at 0.025 % and shows minimum zeta potential before aggregation starts.

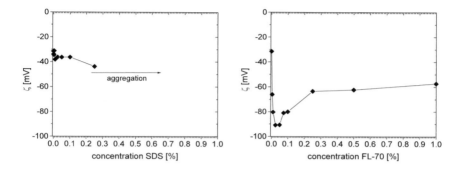

Fig. 4.2. Zeta potentials of natural particles determined as a function of surfactant concentration (from v.d. Kammer (2005))

The clear effect of FL-70 on the zeta potential of the natural reference1 sample in Fig. 4.2 was not always found with natural samples and samples containing iron oxides tended to aggregate under the influence of FL-70. Hence the samples in question were mixed with different carrier compositions prior F^4 analysis and examined after 24h of storage for aggregation or settling.

The main achievement in FFF analysis is the demonstration of MALLS applicability for the analysis of natural samples in FFF, which enabled the on-line evaluation of each FFF fractionation and provides access to particle shape analysis (v.d. Kammer 2005; v.d. Kammer et al. 2005b). While the applicability of FFF-MALLS was shown before for e.g. comparably small latex beads (Thielking et al. 1995) the technique had not been evaluated for natural clay-like particles.

As shown in Fig. 4.3 the simple size distribution of F^4 is supplemented by the on-line particle size determination from MALLS (RMS radius r_g as a function of F^4 retention volume). The right plot in Fig. 4.3 shows that the particle size is linear increasing with retention volume, a proof of ideal fractionation conditions in F^4. However the plot also reveals that non-ideal behaviour can be observed at the beginning and the end of the fractionation. The smallest particles are not fully fractionated from larger ones, most probably due to larger void-peak particles (zero retention volume) eluting among the smaller ones in the fractionation peak. At the end of the fractionation the maximum size of the particles determined is decreasing with increasing cross-flow, an indication of selective losses of larger particles on the membrane in F^4.

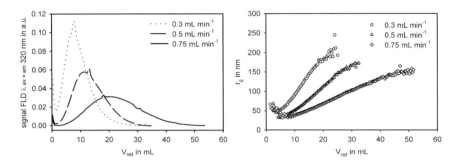

Fig. 4.3. (Left) F^4 fractogramms of reference 5, a variant of reference 1 originating from the same source material, at different constant field settings given as cross-flow rates, detection with Agilent 1100 FLD fluorescence detector in nephelometric mode (v.d. Kammer 2005a). (Right) MALLS determined r_g over retention volume V_{ret} for reference5 at the same cross-flow rates. Determination of r_g: linear Zimm fit, angles 3-18 used (v.d. Kammer 2005)

If the retention volume shown on the x-axis in the right plot in Fig. 4.3 is transformed to hydrodynamic radius via a size calibration it is possible to generate r_g/r_h ratios for each particle size fraction determined by MALLS.

In general the determination of the sphericity of the particles is possible as a function of particle size (deviation from spherical shape) with the coupling of FFF-MALLS. The factor ρ is introduced as the *particle shape factor*. It may be determined by F^4-MALLS analysis in the form of $\rho_s = r_g/r_h$ with r_g the RMS radius as the z-average of particle radius in the close-to monodisperse sample fractions eluting from F^4 and r_h the hydrodynamic radius determined by F^4 (Fig. 4.4). The parameter ρ_v refers to the ratio of the RMS radius r_g determined by MALLS to the volumetric radius r_v as determined by Sedimentation Field-Flow Fractionation (SedFFF: $\rho_v = r_g / r_v$). The two ρ-factors (s and v) take different values for identical non-spherical particles, the ρ_v being the larger in most cases. The reason for this is simply that the r_h already contains a particle shape influence and r_g

and r_h are both increasing with non-sphericity for homogenous particles of identical volume. In contrast the r_v is always shape independent.

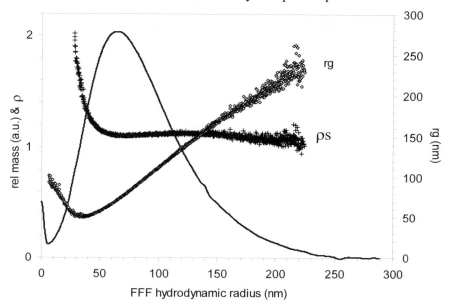

Fig. 4.4. Size distribution (r_h) obtained from F^4-UV/VIS for reference1 (straight line) and calculated MALLS radius of gyration r_g versus rh using a size calibration function for FFF. The factor $r_s = r_g/r_h$ is given as (+). (Reprinted from v.d. Kammer et al. (2005b) with permission from Elsevier)

The combination of SedFFF with quasi on-line PCS measurements provided by the PCS add-on of the *Wyatt Technology Dawn-EOS* MALLS photometer results in the *Perrin* factor F_p for the analysed particles as a function of particle size given by the ratio of r_h over r_v. In principle it is possible to even calculate particle aspect ratios from the three different particle shape factors as long as a geometrical model is applicable for the particles under consideration.

Details of the methodology are presented in v.d. Kammer (2005), v.d. Kammer et al. (2005b).

b) Particle- and Size-Resolved Monitoring of Column Effluents.

Subsequent samples of column effluents from breakthrough and remobilization experiments were analyzed with F^4 to retrieve the size distribution of each sample. The size distributions were arranged to result in three-dimensional time- and particle-size resolved plots.

Dynamic changes in size composition of eluted particles could be shown in the case of laponite, as shown in detail below, but the natural clay suspensions retained their original size distributions in both breakthrough and remobilization experiments.

c) Experimental Setup for Laboratory-Scale Experiments.

For the study of particle transport in soils and filtration units a simplified model system was set up consisting of four glass columns of 2,5 cm inner diameter and 30 cm height, that were filled with pre-cleaned quartz sand (99% purity) of two different grain sizes, fine with 0.22 mm and coarse with 1.6 mm mean grain size. Two columns were run through from base to top under saturated conditions and 2 columns were sprinkled from the top and actively drained at the bottom to provide steady-state unsaturated conditions. Flow rate, carrier composition, composition change in linear gradients, multi-channel sampler, rotation switch to an online analytical chain consisting of 3-angle laser light scattering, UV-VIS, electric conductivity and pH flow-through-measurement devices were all computer-controlled (see Fig. 4.5).

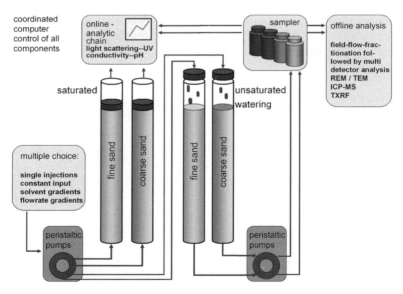

Fig. 4.5. Scheme of the laboratory facility for colloid transport experiments

4.2.3 Laboratory Experiments

One crucial factor that affects the transport of colloids in filtration units and roadside soils is the ionic strength of the transport medium. The retention of colloids at different fixed ionic strengths was monitored with colloid injections in the prepared columns.

The effect of increase and decrease of ionic strength, e. g. besides roads under winter service, was accessed by the use of linear ionic strength gradients of the carrier.

Injections of Clay Mineral Suspensions and Laponite.

The laboratory filtration columns were preconditioned with 1 mmol·L^{-1} sodium-borate / hydrochloric acid buffered carrier at different fixed ionic strengths (1, 10, 15, 35 mmol·L^{-1} Na^+).

At a high flow rate 1 mL of colloid suspension was injected at the inflow of the column and the light scattering signal of the effluent was monitored at 90° in nephelometric mode.

A natural clay suspension (Reference 3, 7.6 mg·L^{-1}) showed complete unretained breakthrough at lowest ionic strength and a linear decrease in recovery for higher ionic strengths, going down to 18.4% at 35 mmol·L^{-1}. Laponite suspension (4 g·L^{-1}) showed a strong increase in UV-VIS as well as light scattering signal intensity between 5 and 10 mmol·L^{-1} Na^+. FFF-analysis of the corresponding effluent samples confirmed the partial formation of agglomerates in the size up to 300 nm. At higher ionic strengths, the recovery rate of laponite is decreasing similar to the natural sample.

Remobilization of particles was simulated by deposition of a injected suspension batch at high ionic strength in the columns. The ionic strength was stepped down to zero, resulting in the release of the retained colloids.

While the polydisperse natural samples show a rapid and complete release without changes in the size distribution, remobilization of Laponite leads to the exclusive elution of agglomerates. The size of these agglomerates changes during the elution from bigger agglomerates at the beginning to smaller ones in the course of the elution (see Fig. 4.6)

Constant Particle Feed and Linear Ionic Strength Gradients.

In a first series of experiments, the stable "illite" suspension was constantly fed into the column along with the carrier at a flow speed of 3 m·d^{-1}, a velocity that is easily approached in filtration media. After an initial equilibration phase of 5 pore volumes, the ionic strength was slowly increased to 25 mmol·l⁻¹ Na⁺ in a linear gradient over 6 days and 67 pore volumes.

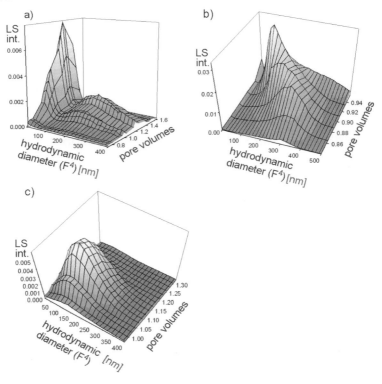

Fig. 4.6. Particle-size- and time- resolved plots: a) Breakthrough of laponite, simultaneous breakthrough of single particles (25-50 nm) and agglomerates (200-300 nm). b) Remobilization of laponite by ionic strength drop, exclusive release of agglomerates. c) Remobilization of a stable soil extract, size distribution matches the unagglomerated feed suspension

At saturated conditions an unretarded breakthrough can be observed after the first pore volume of feeding the suspension. Right at the beginning of the ionic strength gradient, 75% of the particles fed into the column are retained for both saturated and unsaturated fine sand, and 25 % are retained for both saturated and unsaturated coarse sand. Even the presence of a very low ionic strength increases the retention capacity of the columns significantly. In both cases (coarse and fine sand) particles are retained completely on the columns above 10 mmol·L⁻¹ Na⁺. The adsorption capacity of the coarse sand column is exhausted after 60 pore volumes of particle feed, a partial breakthrough is observed even at 25 mmol·L^{-1} ionic strength (see Fig. 4.7).

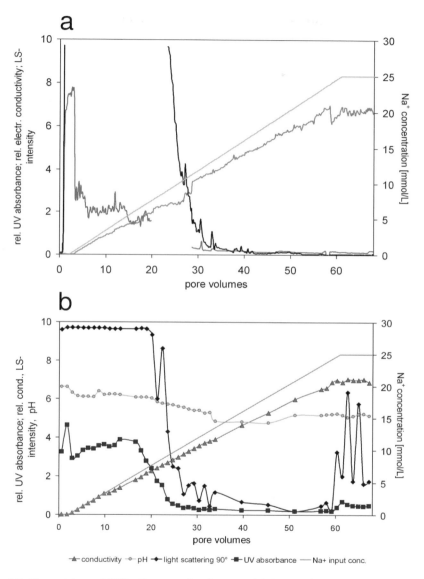

Fig. 4.7. Deposition of "Illite" on sand columns with increasing ionic strength, a) fine sand saturated b) coarse sand unsaturated

The carrier was then switched to a particle-free solution and an ionic strength down-gradient was applied over 60 pore volumes. In all cases, remobilization of the adsorbed colloids began at less than 2 mmol·L^{-1} Na^+ (see Fig. 4.8) and was finished shortly after the ionic strength reached zero. The amount released was nearly equal for all columns except for the unsaturated coarse sand column, that released less colloids due to its limited adsorption capacity.

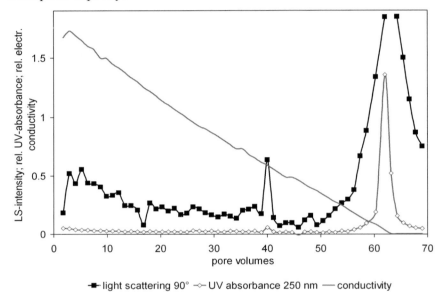

Fig. 4.8. Remobilization of "Illite" in a saturated coarse sand column with an ionic strength down-gradient, the beginning of the remobilization corresponds to a Na^+ - concentration of 2 mmol·L^-

In a second series of experiments the quartz colloid suspension was used for the constant feed experiments. As an initial carrier a solution with 0,01 mmol·L^{-1} Na^+ was used. In contrast to clay mineral suspensions, the first five pore volumes of constant feed of the quartz suspension particles are retained completely on the columns regardless of the grain size of the bed. After that, the eluted particle concentrations rise slowly to reach a full breakthrough only after 20 pore volumes for coarse sand and 35 pore volumes for fine sand.

After a constant particle elution was reached, in linear ionic strength up-gradient was applied over 20 pore volumes to 25 mmol·L^{-1} Na^+, this concentration was kept for a period of 8 pore volumes followed by a down-gradient over 20 pore volumes. Both coarse and fine sand show a complete retention of particles at the beginning of the up-gradient. The remobilization begins during the down-gradient at 6 mmol·L⁻¹ Na⁺, but the intensity of the nephelometric light scattering signal is significantly increased compared to the UV-VIS signal and the intensity at the beginning of the up-gradient. This observation indicates the preferred breakthrough of coarse particles or even agglomerates (see Fig. 4.9).

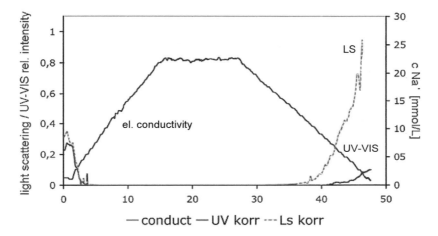

Fig. 4.9. Ionic strength ramp experiment with quartz suspension. The light scattering intensity (Ls korr) at the beginning of the remobilization is much higher than the UV-Vis-signal (UV korr)

Comparison to a Remobilization Experiment with Real Runoff.

A runoff sample collected at the end of a two-week frost period showed a very high concentration of 147 mmol·L⁻¹ sodium. Despite this high ionic strength, 61 mg·L⁻¹ colloids remained in a meta-stable suspension for more than a day. Ten pore volumes of this suspension were fed into a column equilibrated with the same high ionic strength. After one pore volume, a breakthrough of UV/VIS-active substances was observed, followed by a delayed partial breakthrough of particles after two pore volumes. Obviously, part of the suspension remained stabilized while travelling through the column. The supply was then switched to particle-free carrier while retaining the ionic strength.

To simulate the supply of unloaded spring runoff, we applied a down-gradient lowering the ionic strength to zero over 32 pore volumes.

The remobilization breakthrough begins only shortly before the input Na^+ concentration reaches zero, and only small particles with a size distribution from 100 to 500 nm are eluted in a steep release peak (see Fig. 4.10).

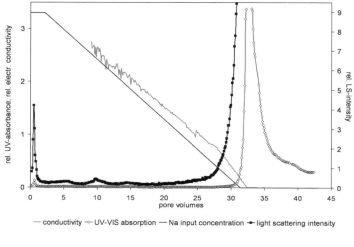

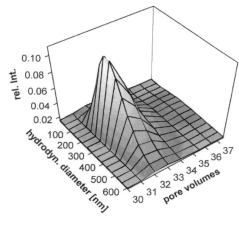

Fig. 4.10. Remobilization of winter road runoff particles retained on a laboratory sand column at high ionic strength

Compared to the model colloids that were examined, the runoff showed different behaviour in many respects.

The fine fraction of suspended matter is not destabilized even at extremely high ionic strengths; a fraction is able to travel through filter matrices without significant retention. However, the release of adsorbed colloids due to the lowering of ionic strength occurs at low Na^+-concentrations beginning approximately at 5 mmol·L^{-1} sodium content for the model colloids as well as for the runoff samples. On-site measurements show a drop in sodium content from 75 mmol·L^{-1} in a winter runoff sample to 0.2 mmol·L^{-1} in a summer runoff sample (see Sect. "Annual runoff variations"). This clearly shows that the drop of ionic strength that occurs at the end of the winter service in natural samples exceeds the critical values for particle remobilization.

For clay minerals and real runoff a retention/release hysteresis can be assumed: Complete retention occurs at a higher ionic strength than remobilization.

4.3 Field Studies

4.3.1 Contamination of the Test Site

The investigated test site is located at a federal road which is frequented by 20,000 vehicles/day half a mile before the driveway to one of the main motorways in Germany. The examined runoff is collected from a road area of 2,000 m^2. During strong rainfall (20 mm in 20 min.; $Q_{eff} = 0.9$), the calculated flow is 30 L·s^{-1} at the outlet of the collecting sewer.

The runoff from six gullies is joined in one tube, drains at the upper part of a hill slope and flows through a meandering channel of 100 m length and a planted filtration stretch into an amphibian-dwelled pond (see Fig. 4.11). Soil samples from different undisturbed location on the site showed high contents of heavy metals in the upper soil layer, especially close to the road (190 mg·kg^{-1} Zn, 175 mg·kg^{-1} Pb, 75 mg·kg^{-1} Cu, 0.8 mg·kg^{-1} Cd).

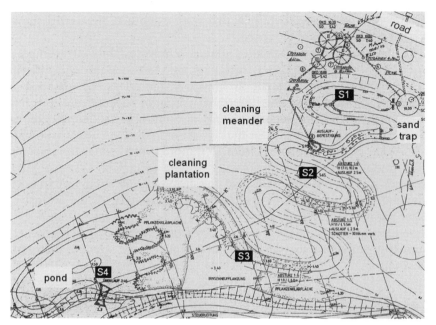

Fig. 4.11. Map of the test site at the B 75 federal highway in Hamburg-Harburg. In the cleaning meander the sampling points for sediments are marked (S1-S4). At the right top the pilot filtration plant with sand trap, splitting well, the two vertical filtration columns and a sampling/collection well is drawn

The sediment of the meander, which consists of deposited runoff particles, shows a high heavy metal load in the upper part which then decreases towards the pond. In the pond sediment, the concentrations are higher again, which indicates its deposit function (see Fig. 4.12). The measured values were compared to the sediment in a "reference" pond nearby, which was built at the same time as the facility, but is only fed by rainwater.

Platinum from car catalyst exhaust is found in the upper soil layer nearest to the road with 80 $\mu g \cdot kg^{-1}$ (F = 200 times background concentration) in the upper part of the meander with 156 $\mu g \cdot kg^{-1}$ (F = 390), and in the pond sediment with 29 $\mu g \cdot kg^{-1}$ (F = 73). Compared to other heavy metals it is less mobile on the discharge path.

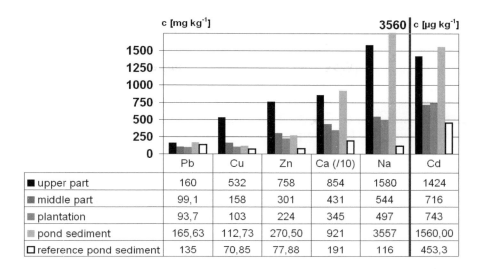

	Pb	Cu	Zn	Ca (/10)	Na	Cd
■ upper part	160	532	758	854	1580	1424
▨ middle part	99,1	158	301	431	544	716
▨ plantation	93,7	103	224	345	497	743
▨ pond sediment	165,63	112,73	270,50	921	3557	1560,00
☐ reference pond sediment	135	70,85	77,88	191	116	453,3

Fig. 4.12. Sediment metal contents in the upper part (S1), middle part (S2) and lower plantation part (S3) of the cleaning meander, as well as in the pond sediment (S4). All metal concentrations except Cd are shown in mg·kg^{-1}, Cd is shown in µg·kg^{-1} for better resolution

4.3.2 Event-Based Characterization of Runoff

Batch runoff samples were taken near the water surface of a 6 m³ sand trap that was installed in addition to the pilot filtration plant described below.

The sampling times were chosen to monitor different types of weather conditions, e. g. strong and weak rain events in different seasons, with a focus on times of frost and snowfall, where the use of deicing salts could be expected. Some of the analytic parameters, like pH value, electric conductivity and oxygen content were measured onsite. The samples were filtered and the amount of solids and the metal contents bound to them were determined and compared to the metal contents in the filtrates.

Annual Runoff Variations

There is a great variation in the solids content of the runoff, which is closely correlated to the actual flow rate at the time of sampling. At stronger rain events more fine particles are washed out of small pits on the road, and additional solids are mobilized due to turbulences in the sand trap. In winter samples, a large fraction of suspended fine particles can be found even at extremely high ionic strengths, as proved by REM analyses, which contradicts the assumption of destabilization of colloidal suspension at high ionic strengths. Part of this phenomenon may be explained by the input of a large amount of very fine particles that served as seeds for the snow crystals with the melting water (see Table 4.2). The metal contents of the solids are not correlated to their amount, but they are very high in the winter samples. The comparably large amount of metals in the winter sample is topped by the metal contents measured in the following thawing period. The total contents of Cu and Zn are five times higher than in the winter sample, other metals are also measured in considerably higher concentrations (see Table 4.3). Actual studies from Sweden (Westerlund and Viklander 2005) confirm these results.

Table 4.2. Examples of the variation of solids and sodium content of six single sampling campaigns at different weather conditions. Additionally the electric conductivity and the measured pH values are depicted

	13.08.02 strong shower	14.10.02 shower	06.01.03 frost snow	13.01.03 thawing rain	20.01.03 weak shower	28.01.03 weak shower
Solids [mg·L^{-1}]	208	18	61	517	75	45
Na$^+$ [mg·L^{-1}]	2		4380	2894	285	25
Ca^{2+} [mg·L^{-1}]	7.37	10.92	83.4	51.1	30.2	8.3
el. cond. [µS·cm^{-1}]	67	73	24400	14400	1780	214
pH	6.0	5.6	6.0	6.0	7.5	5.6

Table 4.3. Total metal contents and those of 0.2 µm-filtrates in a winter and a thawing sample. pH values were at 7.1 for both samples

	[µg·L^{-1}]	Zn	Cu	Cd	Pb	Ni	Cr
06.01.2003	raw water	296	153	1.21	15.8	10.98	11.02
winter	< 0.2 µm	231	23	1.07	0.8	8.25	1.56
13.01.2003	raw water	1721	830	26.01	154.0	42.17	71.02
thawing	< 0.2 µm	288	64	1.28	0.9	6.70	10.41

Sequential filtration of the samples proved to be difficult, since also most of the smaller particles were retained on coarse filter surfaces (5, 1.2

and 0.8 μm pore size). The amount of material retained on coarse filter surfaces was also dependent on the ionic strength. At high sodium contents, more fine particles were able to pass the coarse filters, so the results were not reliable. We concentrated on filtration with 0.2 μm PES-Membranes to separate the complete solid fraction from the solution.

The ratio of dissolved to particle-bound pollutants also shows seasonal differences: The fraction of dissolved metals is much higher in the winter sample than in the thawing sample, nevertheless, there is always a varying amount of metals transported attached to the solids (see Table 4.4).

Table 4.4. Percentage of metals that are bound to runoff particles in a winter (06.01.2003) and thawing (13.01.2003) sample

	Fe	Mn	Zn	Cu	Cd	Pb	Ni	Cr
06.01.2003								
% > 0.2 μm	64%	1%	22%	85%	12%	95%	25%	86%
13.01.2003								
% > 0.2 μm	100%	46%	83%	92%	95%	99%	84%	85%

In general these results show that the solid load and the various percentage of attached pollutants are the main problem for the operation of a filtration plant, since especially at strong rainstorms and thawing events (1) high amounts of runoff occur, (2) big amounts of fine particles are suspended due to increased washing off from the roads and turbulence in subsequent sand traps or retention ponds, (3) a various fraction of the metals is particle-bound, and (4) the decreasing ionic strength during the thawing may lead to the remobilization of particles already deposited on the filter matrix.

The retention capacity of soils which are drained with untreated runoff may also suffer from the combination of these processes.

4.3.3 Pilot Plant for the Clean-up of Runoff

On the test site described above, an additional pilot plant for the clean-up of road runoff was built in 2002, mainly to protect the amphibic fauna in the pond and to test conventional and new filtration materials in this special setting.

The design of the pilot plant is based on the usually limited space available for cleaning units in urban areas. The special circumstances at the site prevented the construction of a typical infiltration ditch design ("Mulden-Rigolen System") as well as an vertical up-stream filter layout. The installed system is based on two down-stream vertical filtration columns of 2 m diameter and 2.5 m usable height that are dimensioned to be able to treat the complete first flush as well as a calculated constant load of 7.6 $l \cdot s^{-1}$ per column at maximum flux with a mean filter grain size of 0.5 mm. To achive maximum flow rates at minimum grain sizes the full available hydraulic head of 2 m had to be used.

Sand Trap

Upstream of the plant the runoff is collected in a sand trap where most of the coarse and sandy materials settle.

At the time when the sand trap was emptied and cleaned during annual maintenance, we took samples of the upper and lower layer of the sand trap sediment that accumulated during one year of operation of the sand trap to analyze its efficiency for the removal of metals.

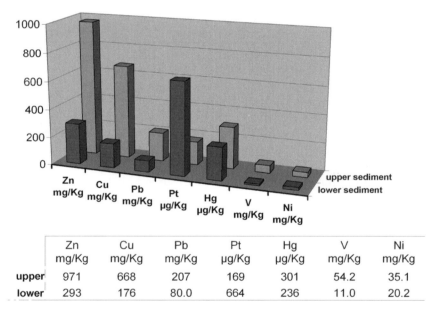

	Zn mg/Kg	Cu mg/Kg	Pb mg/Kg	Pt µg/Kg	Hg µg/Kg	V mg/Kg	Ni mg/Kg
upper	971	668	207	169	301	54.2	35.1
lower	293	176	80.0	664	236	11.0	20.2

Fig. 4.13. Metal contents (based on dry mass) in the upper and lower layer of the sand trap sediment

The sediment is highly contaminated with heavy metals, as shown in Fig. 4.13. Due to the large coarse sand fraction there are less concentrations of heavy metals found in the lower part of the sediment, but Pt and

Hg are accumulated. This can be explained by the exhaust of coarse, Pt-coated pieces of the catalyst matrix that settle rapidly along with the coarse sand fraction.

Operation of the Pilot Plant

In a first step, one of the columns was filled with a multi-layer sand filtration unit: A gravel bed at the bottom of the column with an embedded collecting drainage tube for the cleaned runoff was covered with a layer of polypropylene geotextile. Upon the geotextile are 2 m of pure and calcite-free washed quartz filter sand with a mean grain size of 0.7-1.2 mm. This layer is again covered with a separation layer of geotextile, topped by another gravel drainage layer. The top gravel layer was also covered with a sheet of geotextile to prevent clogging by the large amount of solids still carried along with the water from the sand trap. The filtration column was calculated to retain 90 % of the particle input for the size fraction with the smallest filter factor according to classical filtration theory (Yao et al. 1971, Hofmann 1998).

Cleaning Capacity of the Filter Column

The column was monitored for a year. Due to the geometry of the plant and difficulties with the correct feed flow from the sand trap during operation the total amount of runoff that went through the column could not be quantified.

At the beginning of the operation gravimetric analysis as well as the reduction of Al confirmed the calculated reduction of the particle load. After a year, a significant reduction of the retention capacity for the main metals was observed; the retention capacity for Pb was exhausted, while Zn was still retained with 50 % efficiency (see Table 4.5).

Besides the limited retention capacity of the sand bed, the massive input of small particles from the sand trap proved to be the major problem (as expected from a down-stream design). The clogged top layer of geotextile on the column had to be replaced frequently to secure the flow through the column. After more than a year of operation, this was not enough, because the solids had already passed the layer and clogged deeper areas of the filter.

As a result, the downstream operation of filtration columns with non-reactive filtration material proved to be insufficient for the effective clean-up of road runoff.

Table 4.5. Decreasing percentage of cleaning capacity of the filter column

	23.09.2003	04.02 2004	24.03.2004	21.09.2004
Zn [$\mu g \cdot L^{-1}$]				
before column	225	220	80	130
after column	20	30	5	65
% reduction	91	86	94	50
Cu [$\mu g \cdot L^{-1}$]				
before column	146	125	52	46
after column	15	20	19	35
% reduction	90	84	64	25
Pb [$\mu g \cdot L^{-1}$]				
before column	13	29	10	10
after column	3	1	1	15
% reduction	77	97	90	-release-

Influence of Geotextile Layers

Analysis of the top layer geotextile showed that it trapped big amounts of solids (0.5 to 1 $kg \cdot m^{-1}$) and retained also smaller particles that were trapped in the formation of a filter cake. The metal content of this cake proved to be about three times as high as the concentrations found in the sediment of the sand trap with about 3 $g \cdot kg^{-1}$ dry substance for Zn and Cu as a maximum value (see Table 4.6). In fact the geotextile proved to be very efficient in retaining particulate material.

Table 4.6. Metals retained on the upper geotextile layer

	Cd mg/kg	Zn mg/kg	Cu mg/kg	Pb mg/kg	Ni mg/kg	Hg μg/kg	V mg/kg	Cr mg/kg
GTX 11.12.03	2	1595	1465	401	62	763	95	130
GTX 24.03.04		2723	2802	689	117	1185	203	

4.4 Discussion and Conclusions

The large scale diffuse input of non-degradable pollutants by road runoff poses a major threat to the receiving roadside soils and in some cases the ground waters. Therefore, runoff has to be pre-treated and cleaned before discharge or infiltration. The removal of particles and the large amounts of particle-bound pollutants prior to infiltration in the soil is the key issue of any suitable treatment method. This study shows that runoff possesses special properties that complicate attempts of effective low-cost filtration and will be difficult to deal with. The main problems as well as proposals for advanced filtration techniques are addressed in detail below.

4.4.1 The Importance of Agglomeration Effects

Agglomeration of particles at elevated ionic strength may have significant influence on their retention and release behavior, as shown in the laboratory experiments with laponite. Generally polymorph dispersions are less susceptible to agglomeration, but under favorable conditions at high ionic strength it can also play a role in road runoff, as shown by REM analysis of the suspension at high ionic strength (Fig. 4.14a) and the deposits on the geotextile (Fig. 4.14b).

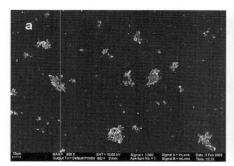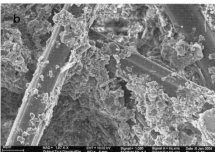

Fig. 4.14. a) Agglomerates filtered from a winter runoff sample. Scale at left lower edge is 10 μm. b) Agglomerated runoff particles on the fiber surfaces of a geotextile filter sheet. Scale at left lower edge is 20 μm

Agglomeration has positive effects on the physical filtration behavior of runoff suspension, f.i. faster sedimentation, retention of smaller particles in filter pores and the restricted diffusion of contaminants from the inside of the aggregate flocs. But on the other hand agglomeration may also have negative influence on the transport of particles. First of all, they reduce the hydraulic conductivity of the filter as soon as they reach sizes larger than the filter pores, small agglomerates may also be transported faster due to size exclusion effects, and the remobilization of contaminated colloids from the break-up of agglomerates may take place at decreasing ionic strength.

4.4.2 Ionic Strength as the Main Problem for Runoff Filtration

Analysis of the runoff showed a great seasonal variability in composition and suspended solids content. At winter service, extremely high ionic strengths could be detected, whereas particle and metal loads were highest at thawing conditions. For an effective filtration, these seasonal effects are the predominant problem. Laboratory experiments showed that the governing factor for the deposition and release of colloids is the ionic strength.

The critical ionic strengths for the deposition and remobilization of particles from filter matrices are low (1-10 mmol·L^{-1} Na^+) but generally higher than the background concentrations of sodium measured during summer periods. A hysteresis effect could be observed in most cases, as ionic strengths for complete particle depositions were higher than those for substantial remobilization.

Apart from that, a fraction of fine particles in road runoff is not susceptible to changes in ionic strength and remains suspended even at high Na^+ concentrations.

Filtration columns operated under unsaturated conditions show less capacity of retaining colloids. The effective pore volume is decreased to the wetted areas, and from the performed experiments there are indications that no measurable amount of colloids was trapped at the water/air interfaces. Additionally, the particle input leads to fast clogging of the filter surfaces.

4.4.3 Physical Factors Influencing Filtration

The grain size of the filter matrices also plays a major role in determining the retention capacity of a filter bed. The results indicate an increase of the retention capacity proportional to the filter surface and a decrease of the grain size, but the hydraulic conductivity suffers from finer filter grains.

The dimension of both filter unit and filter grain size has to be calculated carefully to match the desired filtration capacity.

The operation of a pure sand filter facility under high hydraulic load conditions can not be recommended due to the limited retention capacity and the clogging from the massive input of small particles.

4.4.4 Improved Filtration Methods

The applied geotextile was found to be very effective in retaining large parts of the colloidal fraction. It can also easily be recycled e.g. by back-washing. The arrangement of geotextile sheets in a filtration unit has to prevent early clogging by providing enough effective surface, and it has to be easily accessible to be exchanged or back-flushed.

Filtration materials can be made more effective by (1) adding inexpensive reactive materials such as zeolites, which act as natural ion exchangers and adsorb the solute input fraction of heavy metals and by (2) improving the deposition of colloidal material by controlling the ionic composition and strength of the solution. Divalent cations, such as Ca^{2+}, which can be introduced in small amounts by the use of calcite sand, proved to be very effective for the destabilization of colloidal suspension, as is shown in Laboratory batch experiments. The destabilization efficiency is about 10 times as high as with Na^+.

An upstream operation of filtration columns is essential, because the retention capacity of the filters will only be used fully under constantly saturated conditions, and there is less danger for clogging and formation of filter cakes at the lower surface of the filters. Even the most effective filtration unit will be exhausted after sufficient time of operation, so filters have to be mounted in cartridges that are easily exchangeable.

The high concentrations of de-icing salts in winter run off and the decrease of the ionic strength in the thawing periods will continue to cause problems for all filtration materials that can not be solved at a reasonable price. However, there are alternatives for sodium chloride such as potassium formate, which is biodegraded in the upper soil layers even at low temperature and could be able to eliminate the problem at the source (Hellstén et al. 2005).

References

Allen AG, Nemitz E, Shi JP, Harrison RM, Greenwood JC (2001) Size distributions of trace metals in atmospheric aerosols in the United Kingdom. Atmos Environ 35: 581–4591

Baalousha M, v.d. Kammer F, Motelica-Heino M, Le Coustumer P (2005) Natural sample fractionation by FlFFF-MALLS-TEM: sample stabilization, preparation, pre-Concentration and Fractionation. Journal of Chromatography A, 1093:156-166

Beckett R, Hart B (1993) Use of field-flow fractionation techniques to characterize aquatic particles, colloids, and macromolecules. In: Buffle J, Leeuwen HP (eds.) Environmental Particles Vol.2. IUPAC Environmental analytical and physical chemistry series. Lewis Publishers, Boca Raton, 165-205.

Filella M, Zhang J, Newman M, Buffle J (1997) Analytical applications of photon correlation spectroscopy for size distribution measurements of natural colloidal suspensions: capabilities and limitations. Colloids and Surfaces A, 120:27-46

Förstner U, Jacobs PH, v.d. Kammer F (2001) Impact of natural nanophases on heavy-metal retention in zeolite supported reactive filtration facilities for urban run-off treatment. Fresenius Journal of Analytical Chemistry, 71:652-659

Harrison RM, Tilling R, Romero MSC, Harrad S, Jarvis K (2003) A study of trace metals and polycyclic aromatic hydrocarbons in the roadside environment. Atmos Environ 37:2391–2402

Hassellov M, v.d. Kammer F, Beckett R (2007) Characterization of aquatic colloids and macromolecules by field-flow fractionation. In: Wilkinson K, Lead J (Eds.) Environmental colloids and Particles. IUPAC Series on Analytical and Physical Chemistry of Environmental Systems, 223-276 (in press)

Hellstén PP, Salminen JM, Joergensen KS, Nystén TH (2005) Use of potassium formate in road winter deicing can reduce groundwater deterioration. Environ Sci Technol 39:5095-5100

Hillenbrand Th, Toussaint D, Böhm E, Fuchs S, Scherer U, Rudolphi A, Hoffmann M (2005) Einträge von Kupfer, Zink und Blei in Gewässer und Böden-Analyse der Emissionspfade und möglicher Emissionsminderungsmaßnahmen. Umweltbundesamt Texte 19/2005, ISSN 0722-186x

Hofmann Th. (1998) Kolloidale und suspendierte Partikel: Herkunft, Transport und Relevanz von mobilen Festphasen im Hinblick auf die künstliche Grundwasseranreicherung. Dortmunder Beiträge zur Wasserforschung, 56, 226.

Jacobs PH, v. d. Kammer F, Fitschen T, Förstner U (2000) Filteranlagen auf der Basis natürlichen Clinoptiloliths zur Reinigung kontaminierter Regenwasserabläufe. Jahrestagung der Wasserchemischen Gesellschaft, 29.-31. Mai 2000, Weimar, 399-402.

Kasting U, Gameh O, Grotehusmann D (2001) Bodenfilteranlagen zur Reinigung von Abflüssen stark verschmutzter Verkehrsflächen. Wasserwirtschaft, Abwasser, Abfall 9:1274-1284

Thielking H, Roessner D, Kulicke W-M (1995) On-line coupling of flow field-flow fractionation and multiangle laser light scattering for the characterization of polystyrene particles. Analytical Chemistry, 67:3229-3233.

McCarthy JF, Zachara JM (1989) Subsurface transport of contaminants. Environ Sci Technol 23:496-502

Meißner E (2003) Filteranlagen zur Reinigung von Straßenabflüssen. Berichte aus Wassergüte- und Abfallwirtschaft, Technische Universität München 175, 123-146

Pakkanen TA, Kerminen V-M, Loukkola K, Hillamo RE, Aarnio P, Koskentalo T, Maenhaut W (2003) Size distributions of mass and chemicalcomponents in street-level and rooftop PM1 particles in Helsinki. Atmos Environ 37:1673–1690

Pitt R (ed), Clark S, Parmer K, Field R (1996) Groundwater contamination from stormwater infiltration, Chelsea, Mich. Ann Arbor Press, 1996 ISBN: 1-575-04015-8

Samrani AG, Lartigea BS, Ghanbaja J, Yvon J, Kohler A (2004) Trace element carriers in combined sewer during dry and wet weather: an electron microscope investigation. Water Research, 38:2063–2076

Sternbeck J, Sjödin A, Andreasson K (2002) Metal emissions from road traffic and the influence of resuspension-results from two tunnel studies. Atmos Environ 36:4735–4744

Westerlund C, Viklander M (2005) Particles and associated metals in road runoff during snowmelt and rainfall. Science of the Total Environment (in press)

v.d. Kammer F, Jacobs PH (2003) Dezentrale Regenwasserreinigung an der B75: Problemlösung mit reaktiven Filtersystemen. - In: Wilderer et al. (Eds.): Regenwasserversickerung - eine Möglichkeit dezentraler Regenwasserbewirtschaftung. Berichte aus der Wassergüte- und Abfallwirtschaft, TU München, Vol. 175, pp. 189-203

v.d. Kammer (2005) Characterization of environmental colloids applying field-flow fractionation – multi detection analysis with emphasis on light scattering techniques. PhD thesis at the Hamburg University of Technology, pp 254.

v.d. Kammer F, Baborowski M, Friese K (2005a) Application of HPLC fluorescence detector as a nephelometric turbidity detector following field-flow fractionation to analyse size distributions of environmental colloids. Journal of Chromatography A, 1100:81-89

v.d. Kammer F, Baborowski M, Friese K (2005b) Field-flow fractionation coupled to multi-angle laser light scattering detectors: Applicability and analytical benefits for the analysis of environmental colloids. Analytica Chimica Acta, 552:166-174

v.d. Kammer F, Baborowski M, Tadjiki S, v. Tümpling jr W (2003) Colloidal particles in sediment pore waters: Particle size distributions and associated element-size-distribution in anoxic and re-oxidized samples, obtained by FFF-ICP/MS coupling. Acta hydrochim hydrobiol, 31:400-410

v.d. Kammer F, Förstner U (1998) Natural colloid characterization using flow-field-flow-fractionantion followed by multi-detector analysis. Water Science & Technology, 37:173-180

Yao KM, Habbibian MT, O'Melia CR (1971) Water and wastewater filtration: concepts and applications. Environ Sci Technol 2:1105-1112

Biocolloids and Biofilms

5 Colloid and Microbe Migration in Granular Environments: A Discussion of Modelling Methods

Nathalie Tufenkji

Department of Chemical Engineering, McGill University, Montreal
Quebec H3A 2B2, Canada
Phone: (514) 398-2999
Fax: (514) 398-6678
e-mail: nathalie.tufenkji@mcgill.ca

5.1 Introduction

Predicting the migration of colloids (e.g., clays, manufactured nanomaterials, etc) or biocolloids (e.g., bacteria and viruses) in granular porous media is of significant interest in a number of environmental applications such as granular (deep-bed) filtration used in water and wastewater treatment, riverbank filtration (Medema et al. 2000), *in-situ* bioremediation (Steffan et al. 1999) and protection of groundwater supplies from microbial pathogens and (bio)colloid-associated pollutants (Ferguson et al. 2003). Viruses, protozoa (e.g., *Cryptosporidium*) and certain bacteria are pathogens which may be introduced into the subsurface from various sources such as leaking landfills, septic tanks or land disposal of treated wastewater effluents (Mawdsley et al. 1995; Smith and Perdek 2004). Growing concerns over widespread groundwater contamination and the potential for bioterrorism attacks to the drinking water supply have further prompted the need to develop improved mathematical models of colloid and biocolloid migration in such settings.

The migration of (bio)colloids in granular porous media is typically studied in the laboratory using columns packed with model or natural granular materials. In these studies, the fluid-phase concentration of (bio)colloids in the column effluent is monitored throughout the duration of the experiment. One of the first studies of this kind evaluating microbe migration examined the transport of two bacteriophage in columns packed with natural sediment material (Bales et al. 1989). Since, a substantial research effort has been directed at improving our understanding of the mechanisms controlling the transport and fate of bacteria, viruses and protozoa in granular porous media and these studies are well summarized by Harvey and Harms (Harvey and Harms 2001).

Results of laboratory column experiments and field-scale investigations can be useful for the development of predictive models for (bio)colloid migration in natural and engineered aqueous environments. Mathematical models of (bio)colloid transport in saturated granular porous media generally involve a simplified form of the advection-dispersion equation. In the most commonly used modelling approach, (bio)colloid removal is considered to be mainly controlled by attachment to granular surfaces. In fact, few modelling approaches take into account the influence of the numerous physical, chemical and biological factors which are known to affect the migration of colloids or microorganisms in natural or engineered systems (Yates and Yates 1988). Several reviews have been published which summarize the factors and mechanisms which are believed to influence the migration and deposition of biocolloids in saturated granular media (Yates and Yates 1988; Harvey 1991; Stevik et al. 2004). Yet, the incorporation of this knowledge into mathematical models for predicting the transport and deposition of biocolloids in such systems remains a challenge.

In this chapter, the classical methods used to model colloid and biocolloid migration and deposition in saturated granular media are discussed. The general governing equations commonly considered in models of (bio)colloid transport and fate are presented first. Next, some of the constraints associated to application of these mathematical models are examined (e.g., the difficulties in predicting microbial attachment efficiencies to mineral surfaces). The issues considered in this Chapter are discussed in more detail elsewhere (Tufenkji 2006).

5.2 Classic Methods Used to Model Colloid and Microbe Migration in Granular Porous Media

A number of processes may act independently or concurrently to influence the migration and fate of (bio)colloids in natural and engineered aqueous systems. These mechanisms can be classified as follows: (i) transport processes, (ii) transfer between the liquid phase and the solid phase (due to attachment and detachment), and (iii) inactivation, grazing or death (Tim and Mostaghimi 1991; Tufenkji 2006). Different mathematical approaches have been used to describe these mechanisms in modelling the migration of (bio)colloids in granular porous media and these are presented in this section.

5.2.1 The General Advection-Dispersion Equation

The temporal and spatial variations of (bio)colloid concentration in a homogenous, granular porous medium are described by the advection-dispersion equation (shown here in one spatial dimension for simplicity) (Schijven and Hassanizadeh 2000):

$$\frac{\partial C}{\partial t} = D\frac{\partial^2 C}{\partial x^2} - v\frac{\partial C}{\partial x}.$$ (5.1)

Here, C is the (bio)colloid concentration in the aqueous phase at a distance x and time t, v is the interstitial particle velocity, and D is the hydrodynamic dispersion coefficient. In Eq. 5.1, only the physical transport processes of advection and hydrodynamic dispersion are considered. In granular media, colloids or microorganisms are removed from the fluid-phase by *physicochemical filtration* or attachment to sediment grain surfaces. Physicochemical filtration of microbes has been modeled as either an irreversible (no detachment) or reversible process. In the case of reversible attachment, both equilibrium and kinetic mechanisms have been considered (Yates and Yates 1988). When an attachment mechanism is used to describe removal of particles from the liquid phase, the general equation for (bio)colloid transport and fate in a one-dimensional, homogeneous, granular porous medium becomes (Schijven and Hassanizadeh 2000):

$$\frac{\partial C}{\partial t} + \frac{\rho_b}{\varepsilon}\frac{\partial S}{\partial t} = D\frac{\partial^2 C}{\partial x^2} - v\frac{\partial C}{\partial x},$$ (5.2)

where S is the retained particle concentration, ρ_b is the dry bulk density of the porous medium, and ε is the volumetric water content.

5.2.2 Equilibrium "Adsorption" of Particles

Langmuir and Freundlich isotherms have been used to describe the so-called equilibrium "adsorption" of biocolloids to solid surfaces (Yates and Yates 1988). For the special case of a linear adsorption isotherm, the retained particle concentration is related to the fluid-phase concentration as follows: $S = K_{eq} C$. In this case, Eq. 5.2 can be rewritten as (Schijven and Hassanizadeh 2000):

$$R\frac{\partial C}{\partial t} = D\frac{\partial^2 C}{\partial x^2} - v\frac{\partial C}{\partial x},$$ (5.3)

where $R = 1 + \rho_b K_{eq}/\varepsilon$ is defined as the retardation factor.

5.2.3 Kinetic Attachment and Detachment Processes

An equilibrium adsorption state is not reached instantaneously in a system of (bio)colloids and sediment grains (Schijven and Hassanizadeh 2000). Instead, the removal of particles from the fluid phase is controlled by a kinetic attachment mechanism which consists of two processes. In the first step, (bio)colloids are transferred from the bulk fluid to the surface of the sediment grains (mass transport). In the second step, the particles attach to the surface as a result of physicochemical interactions. The release (detachment) of particles from the grain surface may also be controlled by a kinetic process and the overall rate of change in the concentration of attached particles can be described as follows:

$$\frac{\rho_b}{\varepsilon}\frac{\partial S}{\partial t} = k_{att}C - \frac{\rho_b}{\varepsilon}k_{det}S,$$ (5.4)

where k_{att} and k_{det} are the (bio)colloid attachment and detachment rate coefficients, respectively.

5.2.4 Irreversible Attachment onto Sediment Grains

The classical colloid filtration theory (CFT), which was originally presented by Yao et al (Yao et al. 1971) is the most commonly used approach for evaluating colloid and microbe migration and retention in laboratory and field–scale studies (Harvey and Garabedian 1991; Ryan and Elimelech 1996; Schijven and Hassanizadeh 2000). Indeed, this traditional filtration model is a special case of Eqs. 5.2 and 5.4, whereby the attachment of (bio)colloids to sediment surfaces is considered irreversible, i.e., negligible release of particles (Tufenkji et al. 2003). In the CFT, two parameters are used to describe the overall removal efficiency of particles from the fluid-phase: the single-collector contact efficiency, η_0, reflects the mass transport step and the attachment (collision) efficiency, α, describes the surface attachment step (Yao et al. 1971). Colloid filtration theory is relevant to certain applications of practical interest, where the system may be considered at steady-state and initially free of particles, and the influence of hydrodynamic dispersion is negligible. For this case, the solutions to Eqs. 5.2 and 5.4 subject to the boundary condition $C = C_0|_{x=0}$ for a time period t_0 are (Tufenkji et al. 2003):

$$C(x) = C_0 \exp\left[-\frac{k_{att}}{v} x\right], \qquad (5.5)$$

$$S(x) = \frac{t_0 \varepsilon k_{att} C_0}{\rho_b} \exp\left[-\frac{k_{att}}{v} x\right], \qquad (5.6)$$

where the particle attachment rate coefficient is related to η_0 and α via

$$k_{att} = \frac{3(1-\varepsilon)v}{2d_c} \eta_0 \alpha. \qquad (5.7)$$

Here, d_c is the average sediment grain size. Because available theories are inadequate for predicting the value of the attachment efficiency (Elimelech et al. 1995), researchers use measurements of the normalized fluid-phase colloid or microbe concentration (C/C_0) at a packed length $x=L$ to determine α values for a given system. In this case, the theoretical value of the single-collector contact efficiency (η_0) is determined using a closed-form equation or a computational approach (Yao et al. 1971; Rajagopalan and Tien 1976; Tufenkji and Elimelech 2004a; Nelson and Ginn 2005).

5.2.5 Microbe Inactivation or Die-off

Under certain conditions, it may be important to consider the influence of microbe inactivation or death when predicting the transport and fate of microbes in granular porous media. This irreversible sink mechanism may affect both suspended or attached microorganisms and is commonly described by a first-order rate expression (Corapcioglu and Haridas 1985; Yates and Yates 1988). In natural systems, microbe inactivation may be affected by a number of factors including solution composition and pH, temperature, grazing, and attachment to sediment surfaces. The influence of these factors on the rate of inactivation is not clear however (Yates and Yates 1988).

5.3 Relevance of Classical Modelling Approaches and Proposed Modifications

A number of different approaches have been used to model (bio)colloid migration and removal in granular porous media. As described above, these mathematical models generally differ in the method used to describe the governing particle removal mechanisms. It follows therefore that certain methods may have inherent weaknesses or inconsistencies which should be considered. Some of the limitations of existing modelling approaches and their implications are discussed here. A more detailed discussion is provided elsewhere (Tufenkji 2006).

5.3.1 The Equilibrium Adsorption Mechanism is Considered Inappropriate

Several laboratory and field-scale investigations of biocolloid transport and fate have assumed a so-called equilibrium "adsorption" mechanism to describe particle attachment (Bales et al. 1991; Powelson and Gerba 1994; Chu et al. 2000; Harter et al. 2000; Chen and Strevett 2002; Pang et al. 2003). However, it is important to note that equilibrium adsorption does not result in removal of microorganisms from the fluid phase (Schijven and Hassanizadeh 2000). Rather, this mechanism causes retarded breakthrough of microorganisms in comparison to that of an inert tracer, i.e., it is completely reversible. The retardation factor, R, (defined in Sect. 5.2.2), is a direct measure of this delayed breakthrough, where an R value equal to one implies no retardation. Although several researchers have considered the use of an equilibrium "adsorption" mechanism to describe microbe attachment to sediments, very few studies of microbe migration have reported R values greater than one (Bales et al. 1991; Powelson and Gerba 1994; Dowd et al. 1998; Chen and Strevett 2002). In most studies of microbe migration, little or no particle retardation has been observed (Harvey and Garabedian 1991; Bales et al. 1995; Bales et al. 1997; Ryan et al. 1999; Schijven et al. 1999).

In a field study conducted by Schijven and co-workers (Schijven et al. 1999), comparison of bacteriophage MS-2 transport behavior with that of a salt tracer revealed that phage transport was not retarded by equilibrium adsorption. The slowly rising limb and significant tailing observed in the virus breakthrough curves indicated that virus transport was governed mainly by a kinetic process. Schijven and Hassanizadeh (Schijven and Hassanizadeh 2000) presented numerical results which show that predictions of (bio)colloid transport by an equilibrium approach and a kinetic approach may sometimes lead to similar conclusions. Their calculations demonstrate how the consideration of tailing is required to distinguish between the two mechanisms.

Equilibrium adsorption does not contribute to the removal of particles from the pore space (Schijven and Hassanizadeh 2000). When considering this modelling approach to describe (bio)colloid attachment, the actual removal of particles may be accounted for only by including an additional sink term for attachment and/or an inactivation term (Schijven and Hassanizadeh 2000). Clearly, important inconsistencies can arise from application of an equilibrium sorption model to nonequilibrium conditions (Bales et al. 1991). Consequently, colloid filtration theory (Eq. 5.5), which does not consider equilibrium adsorption, has been applied extensively to evaluate colloid and microbe migration and removal in granular porous media (Bales et al. 1993; Penrod et al. 1996; Pieper et al. 1997; Camesano and Logan 1998; Redman et al. 2001a; Redman et al. 2001b; Abu-Lail and Camesano 2003).

5.3.2 Deviation from CFT in the Presence of Repulsive Electrostatic Interactions

A common approach for evaluating colloid or microbe attachment kinetics is to use colloid filtration theory (Eqs. 5.5 and 5.7) combined with measurements of the fluid-phase particle concentration at the effluent of a packed bed following injection of a particle suspension at the column influent. However, a theoretical analysis of the factors controlling the migration of microorganisms in laboratory-scale columns suggests that the spatial distribution of retained particles along the length of the packed-bed is a more sensitive measure of filtration behavior than the distribution of fluid-phase particles (Tufenkji et al. 2003). Furthermore, well-controlled experimental investigations of (bio)colloid transport and deposition in granular media reveal that the transport and fate of colloids and microbes may not be consistent with CFT (Li et al. 2004; Tufenkji and Elimelech 2004b; Li and Johnson 2005; Tufenkji and Elimelech 2005a; Tufenkji and Elimelech 2005b). Studies have been conducted under controlled physicochemical conditions using uniform spherical latex microspheres and glass bead collectors, where both the fluid-phase particle concentration at the column effluent and the spatial distribution of retained particles have been examined (Li et al. 2004; Li and Johnson 2005; Tufenkji and Elimelech 2005a). These studies have shown that particle deposition behavior is in good agreement with CFT (Eqs. 5.5-5.7) under conditions which are *favorable* for deposition (i.e., in the absence of repulsive electrostatic interactions). However, CFT is generally observed to break down in the presence of repulsive Derjaguin-Landau-Verwey-Overbeek (DLVO) interactions (Derjaguin and Landau 1941; Verwey and Overbeek 1948), namely, under conditions deemed *unfavorable* for particle deposition.

Recent experimental studies conducted with different-sized polystyrene latex microspheres suggest that the observed deviation from CFT may be controlled by the combined influence of different particle deposition mechanisms (e.g., retention of particles in both the secondary and primary energy minima) driven by the presence of heterogeneities in the particle population. To describe the observed particle deposition behavior, a dual deposition mode mechanism has been proposed, whereby a fraction of the (bio)colloid population deposits at a relatively "*slow*" rate in comparison to the remaining particles which deposit at a "*fast*" rate (Tufenkji and Elimelech 2004b). In this case, the colloid or microbe deposition rate coefficient is described using a bimodal distribution. This dual deposition mode model which considers a distribution in the particle deposition rate has also been successfully used to describe the transport and deposition behavior of *Cryptosporidium parvum* (*C. parvum*) oocysts and cells of *Escherichia coli* (*E. coli*) D21. Other laboratory studies involving viruses (Redman et al. 2001a; Redman et al. 2001b) and bacteria (Baygents et al. 1998; Tong et al. 2005) also suggest that a model which takes into account a distribution in deposition rates may be better suited for prediction of microbe migration in granular porous media.

5.3.3 Existing Approaches Fail to Predict the Particle Collision Efficiency, α

Different approaches have been used in an attempt to predict colloid or microbe deposition rates onto collector surfaces in the presence of repulsive electrostatic interactions (i.e., under *unfavorable* deposition conditions). These approaches, which include trajectory analysis for non-Brownian particles (Rajagopalan and Tien 1976), and numerical solution of the convective-diffusion equation (Prieve and Ruckenstein 1974; Elimelech et al. 1995) have been unsuccessful in describing observed particle deposition rates. In general, colloid deposition rates calculated using these models are several orders of magnitude smaller than those determined experimentally. In contrast though, theoretical predictions of colloid (Hahn 1994; Franchi and O'Melia 2003) and microbe (Dong et al. 2002) deposition rates which approach experimental results have been obtained by using a model which considers retention in the secondary energy minimum of the interaction energy profile. A growing body of evidence obtained using different experimental techniques suggests that colloid or microbe retention in a secondary energy well is very likely under conditions where repulsive electrostatic double-layer interactions predominate (Hahn et al. 2004; Hahn and O'Melia 2004; Redman et al. 2004).

Recent efforts to predict colloid and microbe attachment efficiencies on collector surfaces have utilized atomic force microscopy (AFM) derived force-distance measurements between particles and model surfaces (Cail and Hochella 2005b; Cail and Hochella 2005a). Although α values determined from AFM measurements are closer to field data than DLVO-based calculations, the applicability of AFM-derived α values is still quite limited as these are still much smaller than values measured in column or field studies.

Because of the challenges faced in predicting *a priori* the value of colloid or microbe attachment (collision) efficiencies in laboratory or field-scale studies, values of α are evaluated experimentally by considering Eqs. 5.5 and 5.7. To carry out these calculations, a measured or calculated) value of the single-collector contact efficiency, η_0, is required. A new correlation equation for calculating η_0 has recently been developed (Tufenkji and Elimelech 2004a) which overcomes the inherent limitations of previous approaches. The overall single-collector contact efficiency (η_0) is written as the sum of contributions from three major transport mechanisms (Brownian diffusion, interception, and gravitational sedimentation):

$$\eta_0 = 2.4 A_S^{1/3} N_R^{-0.081} N_{Pe}^{-0.715} N_{vdW}^{0.052} + 0.55 A_S N_R^{1.675} N_A^{0.125}$$
$$+ 0.22 N_R^{-0.24} N_G^{1.11} N_{vdW}^{0.053}$$

$$(5.8)$$

The dimensionless parameters, A_S, N_{Pe}, N_A, N_R, N_{vdW}, and N_G, are described in detail in (Tufenkji and Elimelech 2004a). Eq. 5.8 improves upon earlier models by considering the influence of hydrodynamic (viscous) interactions and van der Waals interactions on the deposition of particles by Brownian diffusion. The resulting improvement in predictions of η_0 has direct implications for studies of biocolloid transport, including viruses, bacteria, and protozoa. For instance, calculations based on the Tufenkji and Elimelech (TE) equation indicate that particles in the size range of ~ 2 micrometers (e.g., many bacteria) are nearly twice as mobile in granular porous media than previously thought (Tufenkji and Elimelech 2004a). An example of these calculations is shown in Fig. 5.1, where predictions based on the TE equation are compared to those based on an earlier correlation developed by Rajagopolan and Tien (RT) (Rajagopalan and Tien 1976) as well as a rigorous numerical solution of the convective-diffusion equation. For a 2 μm sized biocolloid, the value of η_0 determined using the TE equation is about 60% smaller than that based on the RT formulation (Fig. 5.1).

Such differences in the calculation of the single-collector contact efficiency can have a marked effect on the prediction of colloid and microbe migration in granular porous media when CFT is applied.

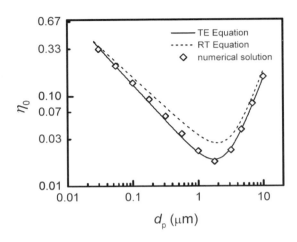

d_p (μm)

Fig. 5.1. Comparison of predictions of single-collector contact efficiency (η_0) based on rigorous numerical solution of the convective-diffusion equation (open diamonds), the RT equation (dashed line), and the TE equation (solid line) for the following conditions: $d_c = 0.40$ mm, $U = 8 \times 10^{-6}$ m/s, $A = 3 \times 10^{-21}$ J, $\varepsilon = 0.36$, $\rho_p = 1.05$ g/cm^3, $T = 288$ K. Values of η_0 are plotted as a function of particle diameter (d_p). Reprinted with permission from (Tufenkji and Elimelech 2004a). Copyright (2004) American Chemical Society.

Eq. 5.8 provides a theoretical prediction for the transport of (bio)colloids to the surface of a spherical collector in granular porous media as a result of Brownian diffusion, interception, and gravitational sedimentation. In certain cases, it may be important to also consider the influence of microbe motility on biocolloid transport (e.g., in the case of motile bacteria propelled by flagella) (Berg 2000). Although few well-controlled studies exist on the role of motility in porous media transport, preliminary results from a theoretical investigation suggest that under certain conditions, cell motility may cause a change in the dependence of η_0 on porewater velocity (Nelson and Ginn 2001). However, it should be noted that this model does not consider the effects of gravitational sedimentation, hydrodynamic (viscous) interactions, and Brownian diffusion on biocolloid attachment. For transport at low flow rates, such as in subsurface natural environments, the two latter mechanisms can be important for biocolloids as large as a few micrometers (Tufenkji and Elimelech 2004a).

5.3.4 Larger Microorganisms May be Affected by Physical Straining

Classical colloid filtration theory does not consider the removal of colloids or microorganisms by physical straining (trapping of particles in pore throats that are too small to allow passage). Under certain conditions of environmental relevance, straining of (bio)colloids may not be an important particle removal mechanism (e.g., physical straining of viruses is unlikely in most granular environments). However, a growing number of experimental studies suggest that straining could be an important particle removal mechanism during porous media transport when the ratio of the (bio)colloid diameter to the median grain diameter (d_p/d_c) is as small as 0.002 (Bradford et al. 2002).

Bradford et al (Bradford et al. 2002; Bradford et al. 2003; Bradford et al. 2005) have been leading current research efforts aimed at better understanding the role of physical straining in colloid and microbe migration in granular porous media. Laboratory-scale particle deposition experiments have been conducted using different-sized model polystyrene microspheres (Bradford et al. 2002) and *Cryptosporidium* oocysts (Bradford and Bettahar 2005) in columns packed with sand grains and/or spherical glass beads. Findings from these studies show that both particle size and grain size have an important effect on the degree of (bio)colloid removal in granular porous media. Fig. 5.2 shows representative breakthrough curves for 2 μm polystyrene latex microspheres introduced into columns packed with different sized quartz sand (Bradford et al. 2002). The coarsest grain size is 710 μm and the finest sand has an average grain size of 150 μm. These experiments show how the degree of particle retention increases considerably with decreasing grain size, implying the important effect of physical particle trapping mechanism.

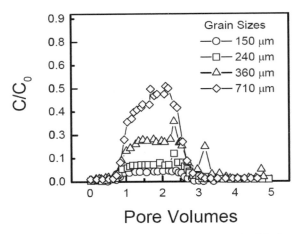

Fig. 5.2. Breakthrough curves for experiments conducted with 2 μm latex microspheres in columns packed with Ottawa sand of decreasing mean grain size (710, 360, 240, and 150 μm). Data reprinted from (Bradford et al. 2002). Copyright (2002) American Geophysical Union.

To describe the results of their experimental work, Bradford et al (Bradford et al. 2003) proposed a model for colloid and microbe transport that considers removal by straining to be an irreversible and depth-dependent mechanism. A comparison of model predictions with experimental results showed that the straining model was generally in better agreement with measurements of effluent colloid and microbe concentration and spatial distributions of retained particles. Bradford and co-workers also suggest that a model which considers particle removal by both attachment and physical straining is more realistic, particularly for systems consisting of intermediate colloid and grain sizes (Bradford et al. 2003). The combined influence of straining and physicochemical filtration on biocolloid removal has also been demonstrated in another study on *Cryptosporidium* transport and deposition (Tufenkji et al. 2004). *Cryptosporidium* transport experiments were conducted in columns packed with glass beads or quartz sand of similar average diameter. This work showed how grain shape can play an important role in determining the straining potential of a granular medium. The straining effect was further investigated by examining the transport of polystyrene latex microspheres of increasing size in deionized (DI) water (Tufenkji et al. 2004). In DI water, the electrostatic double-layer repulsion between microspheres and collectors is substantial so that physicochemical filtration should be negligible. Thus, any removal in the packed column under these conditions can be attributed to the influence of a physical mechanism such as straining. In this experiment, the smallest microspheres (0.32 and 1.0 μm) exhibited no removal after passage in the column packed with quartz grains. The degree of removal for the 1.9 μm particle was also found to be negligible. However, the 4.1 μm particle, which is comparable in size to *C. parvum* oocysts, exhibited considerable removal (~ 40% removal), demonstrating that straining can be an important capture mechanism in this type of granular medium. Furthermore, these researchers observed an increase in oocyst deposition rates with solution salt concentration which confirmed that physicochemical filtration also plays an important role in the removal of *C. parvum* in granular porous media.

5.3.5 Surface Biomolecules Can Enhance or Hinder Microbe Attachment

The surfaces of biocolloids such as bacteria and protozoa are chemically and structurally more complex than most inorganic colloids (Bos et al. 1999). Biomolecules present on cell or (oo)cyst surfaces (e.g., lipopolysaccharides and proteins) can be present in a broad range of sizes, compositions, conformations, and distributions. This inherent heterogeneity of microbe surfaces can give rise to significant difficulties in interpreting and generalizing results of biocolloid transport and attachment studies. Several researchers have attempted to improve our understanding on the role of surface biomolecules in microbial adhesion (Rijnaarts et al. 1996; Williams and Fletcher 1996; Rijnaarts et al. 1999; Considine et al. 2000; Abu-Lail and Camesano 2003; Burks et al. 2003; Kuznar and Elimelech 2005; Kuznar and Elimelech 2006). In general, the results of these studies show that the presence of surface macromolecules can either enhance (Rijnaarts et al. 1995; Rijnaarts et al. 1996; Rijnaarts et al. 1999) or hinder (Rijnaarts et al. 1995; Rijnaarts et al. 1996; Williams and Fletcher 1996; Rijnaarts et al. 1999; Burks et al. 2003) microbe adhesion in aqueous media.

The use of well-characterized model organisms can provide useful insights into the role of cell surface macromolecules on bacterial adhesion. Two groups of researchers (Burks et al. 2003; Walker et al. 2004) used the same three strains of *Escherichia coli* (*E. coli*) K12 to investigate the importance of lipopolysaccharides (LPS) on bacterial migration and attachment. In these studies, the selected organisms differed mainly in LPS length and composition. The results of this work did not reveal any straightforward relationships between the degree of bacterial adhesion and LPS length or microbe surface charge (Burks et al. 2003; Walker et al. 2004). Rather, the researchers proposed that a combination of DLVO-type interactions and surface macromolecule-associated interactions influence bacterial adhesion behavior.

Laboratory experiments can also be designed to examine biocolloid transport and deposition at the microscale, sometimes yielding interesting information on the role of biomolecules in microbe attachment. For instance, studies conducted in a radial stagnation point flow (RSPF) system (Kuznar and Elimelech 2004; Kuznar and Elimelech 2005; Kuznar and Elimelech 2006) and measurements of biocolloid-surface interactions by atomic force microscopy (AFM) (Considine et al. 2000; Considine et al. 2001; Considine et al. 2002) indicate that surface biomolecules can also affect adhesion of protozoa to environmental surfaces. Different research groups have suggested that macromolecules (e.g., proteins) of the *Cryptosporidium parvum* oocyst wall can give rise to a steric repulsion upon approach to a surface in aqueous media, resulting in a lower degree of oocyst attachment. Considine and co-workers used AFM to measure interaction forces between *C. parvum* oocysts and silica over a wide range of environmentally relevant physicochemical conditions (Considine et al. 2000; Considine et al. 2001; Considine et al. 2002). The observed steric interaction between *C. parvum* and a siliceous material was attributed to the presence of a "hairy" protein layer extending from the oocyst surface in a "brushlike" manner. Fig. 5.3 shows the measured normalized force of interaction between an oocyst and a silica surface as a function of separation distance. The thin solid lines represent the DLVO fit for the case of constant potential (lower limit) and constant charge (upper limit). The thicker solid line corresponds to the force predicted by the theory of Pincus for surfaces bearing a polyelectrolyte brush layer. This latter calculation is based on a brush width of 50 nm. Comparison of the measured force curve with model calculations suggests that the oocyst surface can be described as a polyelectrolyte brush at intermediate separation distances (Considine et al. 2001).

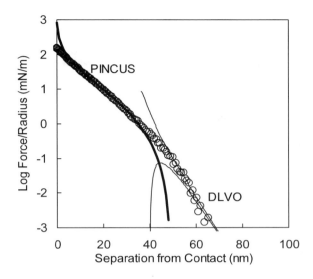

Fig. 5.3. Normalized force of interaction between an oocyst and a silica sphere (open symbols) in 1 mM KNO_3 at pH 8.9. The thin solid lines correspond to the result of the DLVO calculation based on the corresponding zeta-potentials and the bulk electrolyte concentration, with the DLVO plane of charge origin being shifted around 35 nm to obtain an order of magnitude fit. The thicker solid line corresponds to the result of the force predicted from a Pincus model of a polyelectrolyte brush of width 50 nm. Reprinted with permission from (Considine et al. 2001). Copyright (2001) American Chemical Society.

Kuznar and Elimelech (Kuznar and Elimelech 2005) provided further evidence in support of this theory by comparing the deposition behavior of viable oocysts with those of inactivated oocysts (both formalin and heat treated). Low deposition rates are measured with viable oocysts even at high solution ionic strengths when electrostatic energy barriers are eliminated. This behavior is attributed to a steric repulsion between viable *C. parvum* oocysts and the quartz surface. Oocyst deposition rates increased significantly after oocysts were inactivated with formalin or heat. Kuznar and Elimelech (Kuznar and Elimelech 2005) linked the increased deposition rates to postulated changes in the structure of oocyst surface proteins caused by formalin and heat treatments. Similar experiments have been conducted with viable *C. parvum* oocysts and oocysts treated with the digestive enzyme proteinase K for removal of oocyst wall proteins (Kuznar and Elimelech 2006). In this study, treatment of proteinase K resulted in significantly higher oocyst deposition rates onto a quartz surface. The authors attributed this behavior to an "electrosteric" repulsion between surface macromolecules of the untreated oocyst and the quartz surface which was eliminated upon cleavage of the wall proteins (Kuznar and Elimelech 2006).

Researchers are only just beginning to have an understanding of the role of surface biomolecules in biocolloid migration and retention in granular porous media. When a microorganism approaches a surface in an aqueous environment, the presence of macromolecules such as LPS or proteins on cell/(oo)cyst surfaces can give rise to a broad range of interactions. To date, we do not have a sufficient understanding of these interactions which allows us to consider related attachment and detachment mechanisms in biocolloid migration and retention models.

5.3.6 Growth and Death Rates of Microorganisms Are Highly Variable

Careful inspection of Eqs. 5.5-5.7 reveals that colloid filtration theory (CFT) does not consider the influence of microbe growth nor death on the overall transport behavior of microorganisms. In general, this can be attributed to the fact that microbial growth and death rates in granular aqueous environments are not well understood. Specifically, there exists very little useful information on inactivation rates and growth rates for microorganisms attached to sediment surfaces. More studies have been conducted to examine microbe inactivation rates in aqueous media, and these are typically represented using first-order rate expressions (Corapcioglu and Haridas 1985; Schijven and Hassanizadeh 2000).

Bacterial growth may be described using the well-known Monod relationship, however its relevance to subsurface environments has not been demonstrated (Corapcioglu and Haridas 1985; Harvey 1991). One challenge associated with this approach is predicting the values of the Monod parameters, namely, the maximum growth rate and the affinity constant. These microbe-specific "constants" are influenced by changes in environmental conditions and substrate. Furthermore, it is not easy to obtain reliable estimates of these parameters for *in-situ* conditions because of the difficulty in obtaining uncontaminated samples and accounting for temporal and spatial variations in microbial activity. The development and use of novel imaging devices and flow chambers is likely to improve predictions of microbial growth and inactivation rates in these environments (Dupin and McCarty 1999; Yarwood et al. 2002).

Microbe survival in granular environments is influenced by several factors, such as the type of organism, pore water chemistry and temperature, the presence of predators and parasites, and degree of attachment (Yates and Yates 1988; Harvey 1991). Interaction amongst some of these factors is also possible and presents an additional challenge when determining the role of inactivation (or death) in biocolloid migration and retention. Several studies have attempted to elucidate the influence of various environmental factors on the inactivation of viruses (Yates and Yates 1988; Schijven and Hassanizadeh 2000; Ryan et al. 2002; Schijven et al. 2002). This work demonstrates how complicated it is to draw any general conclusions on the role of certain factors in virus inactivation. For example, the effect of attachment on virus inactivation is not clear - attachment to some geological surfaces accelerates inactivation, whereas attachment to other surfaces may protect viruses from inactivation (Schijven and Hassanizadeh 2000; Ryan et al. 2002; Harvey and Ryan 2004). This variability in biocolloid growth and inactivation processes suggests that incorporation of these mechanisms into predictive models of microbe migration should be treated on a case-by-case basis.

5.4 Concluding Remarks

A number of different mathematical approaches have been applied to model the behavior of colloids and microorganisms in natural and engineered aqueous granular environments. Some of these modelling methods have inherent weaknesses or limitations which are often overlooked or ignored by researchers. For example, the relevance of colloid filtration theory seems to be limited mainly to ideal systems where conditions which are *favorable* for deposition predominate. Some researchers suggest modifications of classic transport models such as consideration of a distribution in (bio)colloid deposition rates or a depth-dependent straining function. In the long-term, these modifications to classic modelling approaches should provide improvements to predictions of microbe transport and retention in granular media. However, a great deal more research is necessary to improve our understanding of (i) the role of surface biomolecules on biocolloid-surface interactions, (ii) physical straining of biocolloids in granular systems, and (iii) microbe growth and inactivation rates when retained on sediment surfaces. More studies are also required to develop improved techniques to measure the value of biocolloid attachment efficiencies onto sediment surfaces. Atomic force microscopy may prove to be very useful in this respect. A better understanding of fundamental biocolloid-surface interactions will allow us to develop more comprehensive models which consider the broad range of microbe transport and deposition behavior encountered in complex natural environments.

Acknowledgements

The author acknowledges the support of the Brace Centre for Water Resources Management at McGill University and the Canada Research Chairs Program.

Notation

Symbols

A_S	porosity-dependent parameter of Happel's model
C	fluid-phase (suspended) particle concentration
C_0	bulk (influent) particle concentration
D	hydrodynamic dispersion coefficient
d_c	diameter of spherical collector
d_p	diameter of particle
K_{eq}	equilibrium distribution coefficient
k_{att}	particle attachment rate coefficient
k_{det}	particle detachment rate coefficient
N_A	attraction number
N_G	gravity number
N_{Pe}	Peclet number
N_R	aspect ratio
N_{vdW}	van der Waals number
R	retardation factor, $R = 1 + \rho_b K_{eq}/\varepsilon$
S	attached particle concentration
T	absolute temperature
t	time
t_0	duration of particle injection
U	approach (superficial) velocity of fluid
v	interstitial pore velocity; $v = U/\varepsilon$
x	distance

Greek Symbols

α	particle attachment (collision) efficiency
ε	porosity of a porous medium
η_0	overall single-collector contact efficiency
ρ_b	dry bulk density of porous medium
ρ_p	particle density

References

Abu-Lail NI, Camesano TA (2003) The role of lipopolysaccharides in the adhesion, retention, and transport of *Escherichia coli* JM109. Environ Sci Technol 37:2173-2183

Bales RC, Gerba CP, et al. (1989) Bacteriophage transport in sandy soil and fractured tuff. Applied and Environmental Microbiology 55 (8):2061-2067

Bales RC, Hinkle SR, et al. (1991) Bacteriophage adsorption during transport through porous media: chemical perturbations and reversibility. Environ Sci Technol 25 (12):2088-2095

Bales RC, Li S, et al. (1993) MS-2 and poliovirus transport in porous media: hydrophobic effects and chemical perturbations. Water Resources Research 29 (4):957-963

Bales RC, Li S, et al. (1995) Virus and bacteria transport in a sandy aquifer, Cape Cod, MA. Ground Water 33 (4):653-661

Bales RC, Li S, et al. (1997) Bacteriophage and microsphere transport in saturated porous media: forced-gradient experiment at Borden, Ontario. Water Resources Research 33 (4):639-648

Baygents JC, Glynn JR, et al. (1998) Variation of surface charge density in monoclonal bacterial populations: Implications for transport through porous media. Environ Sci Technol 32 (11):1596-1603

Berg, HC (2000) Motile behavior of bacteria. Physics Today 53 (1):24-29

Bos R, van der Mei HC, et al. (1999) Physico-chemistry of initial microbial adhesive interactions - its mechanisms and methods for study. FEMS Microbiology Reviews 23 (2):179-230

Bradford SA, Bettahar M (2005) Straining, attachment, and detachment of *Cryptosporidium* Oocysts in saturated porous media. Journal of Environmental Quality 34:469-478

Bradford SA, Simunek J, et al. (2005) Straining of colloids at textural interfaces. Water Resources Research 41:W10404, doi:10.1029/2004WR003675

Bradford SA, Simunek, J et al. (2003) Modeling colloid attachment, straining, and exclusion in saturated porous media. Environ Sci Technol 37:2242-2250

Bradford SA, Yates SR, et al. (2002) Physical factors affecting the transport and fate of colloids in saturated porous media. Water Resources Research 38 (12):1327-1338

Burks GA, Velegol SB, et al. (2003) Macroscopic and nanoscale measurements of the adhesion of bacteria with varying outer layer surface composition. Langmuir 19:2366-2371

Cail TL, Hochella MF (2005a) The effects of solution chemistry on the sticking efficiencies of viable *Enterococcus faecalis*: An atomic force microscopy and modeling study. Geochmica et Cosmochimica Acta 69 (12):2959-2969

Cail TL, Hochella MF (2005b) Experimentally derived sticking efficiencies of microparticles using atomic force microscopy. Environ Sci Technol 39:1011-1017

Camesano TA, Logan BE (1998) Influence of fluid velocity and cell concentration on the transport of motile and nonmotile bacteria in porous media. Environ Sci Technol 32:1699-1708

Chen G, Strevett K (2002) Surface free energy relationships used to evaluate microbial transport. Journal of Environmental Engineering 128 (5):408-415

Chu Y, Jin Y et al. (2000) Virus transport through saturated sand columns as affected by different buffer solutions. Journal of Environmental Quality 29:1103-1110

Considine RF, Dixon DR et al. (2000) Laterally-resolved force microscopy of biological microspheres - Oocysts of *Cryptosporidium parvum*. Langmuir 16:1323-1330

Considine RF, Dixon DR et al. (2002) Oocysts of *Cryptosporidium parvum* and model sand surfaces in aqueous solutions: an atomic force microscope (AFM) study. Water Research 36:3421-3428

Considine RF, Drummond CJ et al. (2001) Force of interaction between a biocolloid and an inorganic oxide: Complexity of surface deformation, roughness, and brushlike behavior. Langmuir 17:6325-6335

Corapcioglu MY, Haridas A (1985) Microbial transport in soils and groundwater: a numerical model. Advances in Water Resources 8:188-200

Derjaguin BV, Landau LD (1941) Theory of the stability of strongly charged lyophobic sols and of the adhesion of strongly charged particles in solutions of electrolytes. Acta Physicochim. URSS 14:733-762

Dong H, Onstott TC et al. (2002) Theoretical prediction of collision efficiency between adhesion-deficient bacteria and sediment grain surface. Colloids and Surfaces B-Biointerfaces 24:229-245

Dowd DE, Pillai SD et al. (1998) Delineating the specific influence of virus isoelectric point and size on virus adsorption and transport through sandy soils. Applied and Environmental Microbiology 64(2):405-410

Dupin HJ, McCarty PL (1999) Mesoscale and microscale observations of biological growth in a silicon pore imaging element. Environ Sci Technol 33:1230-1236

Elimelech M, Gregory J et al. (1995) Particle deposition and aggregation: measurement, modelling, and simulation. Oxford, England, Butterworth-Heinemann

Ferguson C, de Roda Husman AM et al. (2003) Fate and transport of surface water pathogens in watersheds. Critical Reviews in Environmental Science and Technology 33 (3):299-361

Franchi A, O'Melia CR (2003) Effects of natural organic matter and solution chemistry on the deposition and reentrainment of colloids in porous media. Environ Sci Technol 37:1122-1129

Hahn MW (1994) Ph.D. Thesis. Baltimore, MD., The Johns Hopkins University

Hahn MW, Abadzic D et al. (2004) Aquasols: On the role of secondary minima. Environ Sci Technol 38:5915-5924

Hahn MW, O'Melia CR (2004) Deposition and reentrainment of Brownian particles in porous media under unfavorable chemical conditions: Some concepts and applications. Environ Sci Technol 38:210-220

Harter T, Wagner S et al. (2000) Colloid transport and filtration of *Cryptosporidium* parvum in sandy soils and aquifer sediments. Environ Sci Technol 34 (1):62-70

Harvey RW (1991) Parameters involved in modeling movement of bacteria in groundwater. Modeling the environmental fate of microorganisms. Hurst CJ. Washington, D. C., American Society for Microbiology 89-114

Harvey RW, Garabedian SP (1991) Use of colloid filtration theory in modeling movement of bacteria through a contaminated sandy aquifer. Environ Sci Technol 25 (1):178-185

Harvey RW, Harms H (2001) Transport of microorganisms in the terrestrial subsurface: In situ and laboratory methods. Manual of Environmental Microbiology. Hurst CJ. Washington, D.C., ASM Press 753-776

Harvey RW, Ryan JN (2004) Use of PRD1 bacteriophage in groundwater viral transport, inactivation, and attachment studies. FEMS Microbiology Ecology 49:3-16

Kuznar ZA, Elimelech M (2004) Adhesion kinetics of viable *Cryptosporidium parvum* oocysts to quartz surfaces. Environ Sci Technol 38 (24):6839-6845

Kuznar ZA, Elimelech M (2005) Role of surface proteins in the deposition kinetics of *Cryptosporidium parvum* oocysts. Langmuir 21:710-716

Kuznar ZA, Elimelech M (2006) *Cryptosporidium* oocyst surface macromolecules significantly hinder oocyst attachment. Environ Sci Technol 40:1837-1842

Li X, Johnson WP (2005) Nonmonotonic variations in deposition rate coefficients of microspheres in porous media under unfavorable deposition conditions. Environ Sci Technol 39:1658-1665

Li X, Scheibe TD et al. (2004) Apparent decreases in colloid deposition rate coefficients with distance of transport under unfavorable deposition conditions: A general phenomenon. Environ Sci Technol 38:5616-5625

Mawdsley JL, Bardgett RD et al. (1995) Pathogens in livestock waste. Their potential for movement through soil and environmental-pollution. Applied Soil Ecology 2 (1):1-15

Medema GJ, Juhasz-Holterman MHA et al. (2000) Removal of micro-organisms by bank filtration in a gravel-sand soil. Proc. of the Int. Riverbank Filtration Conference. Julich W, Schubert J. Düsseldorf, Germany, IAWR

Nelson KE, Ginn TR (2001) Theoretical investigation of bacterial chemotaxis in porous media. Langmuir 17:5636-5645

Nelson KE, Ginn TR (2005) Colloid filtration theory and the Happel sphere-in-cell model revisited with direct numerical simulation of colloids. Langmuir 21:2173-2184

Pang L, Close M et al. (2003) Estimation of septic tank setback distances based on transport of E. coli and F-RNA phages. Environment International 29:907-921

Penrod SL, Olson TM et al. (1996) Deposition kinetics of two viruses in packed beds of quartz granular media. Langmuir 12 (23):5576-5587

Pieper AP, Ryan JN et al. (1997) Transport and recovery of bacteriophage PRD1 in a sand and gravel aquifer: effect of sewage-derived organic matter. Environ Sci Technol 31:1163-1170

Powelson DK, Gerba CP (1994) Virus removal from sewage effluents during saturated and unsaturated flow through soil columns. Water Research 28 (10):2175-2181

Prieve DC, Ruckenstein E (1974) Effect of London Forces upon rate of deposition of Brownian particles. AIChE Journal 20 (6):1178-1187

Rajagopalan R, Tien C (1976) Trajectory analysis of deep-bed filtration with sphere-in-cell porous media model. Aiche Journal 22 (3):523-533

Redman JA, Estes MK et al. (2001a) Resolving macroscale and microscale heterogeneity in virus filtration. Colloids and Surfaces A: Physicochemical and Engineering Aspects 191 (1-2):57-70

Redman JA, Grant SB et al. (2001b) Pathogen filtration, heterogeneity, and the potable reuse of wastewater. Environ Sci Technol 35 (9):1798-1805

Redman JA, Walker SL et al. (2004) Bacterial adhesion and transport in porous media: role of the secondary energy minimum. Environ Sci Technol 38:1777-1785

Rijnaarts HHM, Norde W et al. (1995) Reversibility and mechanism of bacterial adhesion. Colloids and Surfaces B-Biointerfaces 4:5-22

Rijnaarts HHM, Norde W et al. (1996) Bacterial deposition in porous media related to the clean bed collision efficiency and to substratum blocking by attached cells. Environ Sci Technol 30 (10):2869-2876

Rijnaarts HHM, Norde W et al. (1999) DLVO and steric contributions to bacterial deposition in media of different ionic strengths. Colloids and Surfaces B-Biointerfaces 14 (1-4):179-195

Ryan JN, Elimelech M (1996) Colloid mobilization and transport in groundwater. colloids and surfaces A: Physicochemical and Engineering Aspects 107:1-56

Ryan JN, Elimelech M et al. (1999) Bacteriophage PRD1 and silica colloid transport and recovery in an iron oxide-coated sand aquifer. Environ Sci Technol 33 (1):63-73

Ryan JN, Harvey RW et al. (2002) Field and laboratory investigations of inactivation of viruses (PRD1 and MS2) attached to iron oxide-coated quartz sand. Environ Sci Technol 36:2403-2413

Schijven JF, Hassanizadeh SM (2000) Removal of viruses by soil passage: Overview of modeling, processes, and parameters. Critical Reviews in Environmental Science and Technology 30 (1):49-127

Schijven JF, Hassanizadeh SM et al. (2002) Column experiments to study nonlinear removal of bacteriophages by passage through saturated dune sand. Journal of Contaminant Hydrology 58:243-259

Schijven JF, Hoogenboezem W et al. (1999) Modeling removal of bacteriophages MS2 and PRD1 by dune recharge at Castricum, Netherlands. Water Resources Research 35 (4):1101-1111

Smith JE, Perdek JM (2004) Assessment and management of watershed microbial contaminants. Critical Reviews in Environmental Science and Technology 34:109-139

Steffan RJ, Sperry KL et al. (1999) Field-scale evaluation of in situ bioaugmentation for remediation of chlorinated solvents in groundwater. Environ Sci Technol 33:2771-2781

Stevik TK, Aa K et al. (2004) Retention and removal of pathogenic bacteria in wastewater percolating through porous media: a review. Water Research 38:1355-1367

Tim US, Mostaghimi S (1991) Model for predicting virus movement through soils. Ground Water 29 (2):251-259

Tong MP, Camesano TA et al. (2005) Spatial variation in deposition rate coefficients of an adhesion-deficient bacterial strain in quartz sand. Environ Sci Technol 39:3679-3687

Tufenkji N (2006) Modeling microbial transport in porous media: Traditional approaches and recent developments. Advances in Water Resources doi:10.1016/j.advwatres.2006.05.014.

Tufenkji N, Elimelech M (2004a) Correlation equation for predicting single-collector efficiency in physicochemical filtration in saturated porous media. Environ Sci Technol 38:529-536

Tufenkji N, Elimelech M (2004b) Deviation from the classical colloid filtration theory in the presence of repulsive DLVO interactions. Langmuir 20:10818-10828

Tufenkji N, Elimelech M (2005a) Breakdown of colloid filtration theory: Role of secondary energy minimum and surface charge heterogeneities. Langmuir 21:841-852

Tufenkji N, Elimelech M (2005b) Spatial distributions of *Cryptosporidium* oocysts in porous media: Evidence for dual mode deposition. Environ Sci Technol 39:3620-3629

Tufenkji N, Miller GF et al. (2004) Transport of *Cryptosporidium* oocysts in porous media: Role of straining and physicochemical filtration. Environ Sci Technol 38:5932-5938

Tufenkji N, Redman JA et al. (2003) Interpreting deposition patterns of microbial particles in laboratory-scale column experiments. Environ Sci Technol 37:616-623

Verwey EJW, Overbeek JTG (1948) Theory of the stability of lyophobic colloids. Amsterdam, Elsevier

Walker SL, Redman JA et al. (2004) Role of cell surface lipopolysaccharides (LPS) in *Escherichia coli* K12 Adhesion and transport. Langmuir 20:7736-7746

Williams V, Fletcher M (1996) *Pseudomonas fluorescens* adhesion and transport through porous media are affected by lipopolysaccharide composition. Applied and Environmental Microbiology 62 (1):100-104

Yao KM, Habibian MT et al. (1971) Water and waste water filtration - Concepts and applications. Environ Sci Technol 5 (11):1105-1112

Yarwood RR, Rockhold ML et al. (2002) Noninvasive quantitative measurement of bacterial growth in porous media under unsaturated-flow conditions. Applied and Environmental Microbiology 68 (7):3597-3605

Yates MV, Yates SR (1988) Modeling microbial fate in the subsurface environment. Critical Reviews in Environmental Control 17 (4):307-344

6 Influence of Biofilms on Colloid Mobility in the Subsurface

Martin Strathmann, Carlos Felipe Leon-Morales, Hans-Curt Flemming

Biofilm Centre, University of Duisburg-Essen, Geibelstrasse 41, D-47057
Duisburg, Germany,
e-mail: HansCurtFlemming@compuserve.com

6.1 Colloids and Biofilms - an Introduction

Transport processes in subsurface environments are determined by complex interactions between the soil matrix and dissolved as well as particulate substances. Biofilms play an important role in the transport of colloids in the subsurface, since biofilms cover the solid soil matrix and hence influence the interaction of colloids with the soil matrix. Consequently, biofilms can influence the mobility of colloids and colloid-bound contaminants either by deposition of colloids within the biofilm matrix, by remobilization of bound colloids, and/or by co-elution of colloids together with detaching biofilm compartments. Further, biofilm organisms can take part in the degradation of colloids or colloid-bound contaminants as well as in colloid generation processes.

Biofilms are involved in interactions between dissolved or particulate matter and solid phases, because they colonize and cover large areas of wetted surfaces. They constitute a gel phase, which is involved in all sorption processes and possesses different sorption mechanisms and kinetics than e. g. minerals. This fact has not been taken into account in most studies on colloid transport so far. A large amount of contaminants in rain drainage water for example from roof or road runoff is mobilized by the water phase in colloidal or colloid-bound form and not in a dissolved state. This has been investigated in the past only to a very limited extent.

In 2002 an international workshop on "Colloids and colloid-facilitated transport of contaminants in soil and sediments" was held in Denmark. The general consensus was that the soil liquid and gaseous phases were the only mobile phases that could facilitate transport in the soil matrix. For about two decades, however, it has been suspected the influence of solid mobile phases in contaminant transport. It is now generally accepted that also a part of the solid phase is mobile in the form of colloids and can facilitate chemical transport (De Jonge et al. 2004). Nevertheless, the role of microorganisms and biofilms in the mobility of pollutants and colloids in the subsurface was not recognized at the time. More recently, the influence of biofilms on the mobility of colloids was acknowledged at the international workshop "Colloids and colloid mediated transport of contaminants in soils" held in 2005 in Bad Herrenalb (Germany).

6.2 Biofilms in the Subsurface

6.2.1 What Is a Biofilm?

Most microorganisms live and grow in aggregates such as films on interfaces (biofilms in the strict sense), flocs (floating biofilms), and sludge. This form of microbial life is subsumed under the expression "biofilm". The feature which is common to all these forms of aggregated microbial life is that the microorganisms are embedded in a matrix of extracellular polymeric substances (EPS) which are responsible for the morphology, physico-chemical properties, structure, coherence, and ecological features of these aggregates (Flemming and Wingender 2002). A modern definition of biofilms was given by Donlan and Costerton (2002): "A biofilm is a microbially derived sessile community characterized by cells that are irreversibly attached to a substratum or interface or to each other, are embedded in a matrix of EPS that they have produced, and exhibit an altered phenotype with respect to growth rate and gene transcription".

The formation of biofilms starts with the first contact of a microorganism to a surface. Surfaces immersed into water become, within seconds, covered with a so-called conditioning film, which consists of macromolecules such as humic substances, polysaccharides and proteins that are present in trace amounts in water. The organisms adhere to a surface for a certain period of time until they begin to multiply. Multiplication is determined in this phase by nutrient availability, and not by the cell number in suspension. In most systems, biofilm growth will be limited either by nutrient scarcity or by shear forces, and remain in a plateau phase with large fluctuations when parts of the biofilm are detached. In natural environments and technical systems, biofilms occur mostly in the plateau phase in equilibrium between external mechanical stress and internal mechanical stability. The level of plateau is subject to large fluctuations due to numerous unpredictable events like e.g. changes in nutrient availability, population shifts, grazing, gas formation, variations in sheer forces, and irregular sloughing events.

Although biofilms are formed and inhabited by individual microbial cells, they form highly structured microenvironments and exhibit complex interactions and thus can be considered as an approach of nature to multicellular organisms. This has for example been described for the coordinated multicellular behaviour during the mixed development of *Escherichia coli* and *Bacillus subtilis* colonies, the spatially organized interspecies metabolic cooperation in anaerobic bioreactor granules (Shapiro 1998), and the migration of populations of *Serratia liquefaciens* by swarming motility (Dworkin 1996). The organisms benefit from multicellular cooperation by using cellular division of labour, accessing resources that cannot effectively be utilized by single cells and optimizing population survival by differentiation into distinct phenotypes. The biofilm matrix allows for the establishment of microcolonies of different species, which can develop synergistic interactions. Wolfaardt et al. (1994) showed that such microconsortia were able to degrade complex substrates, which could not be utilized by single members of the community. Watnick and Kolter (2000) called biofilms a "city of microbes", referring to the many functional cooperative and adverse aspects provided by the matrix in which the biofilm organisms dwell.

6.2.2 Extracellular Polymeric Substances

As mentioned above, the microorganisms within biofilms are embedded in a slimy matrix of EPS. EPS were defined as "extracellular polymeric substances of biological origin that participate in the formation of microbial aggregates" (Geesey 1982). Nevertheless, some researches misuse the abbreviation EPS for exopolysaccharides. More recently, a more general definition for EPS was given by Wingender et al. (1999a) as a comprehensive term for different classes of organic macromolecules which occur in the intercellular spaces of microbial aggregates.

The EPS represent the construction material of biofilms. A detailed review about EPS can be found in Wingender et al. (1999a) and Flemming and Wingender (2002). EPS are mainly responsible for the structural and functional integrity of biofilms and are considered as the key components that determine the physico-chemical and biological properties of biofilms (Allison 2003; Flemming and Wingender 2003b; 2003a). EPS form a three-dimensional, gel-like, highly hydrated and often charged matrix in which the microorganisms are more or less immobilized in close proximity to each other. EPS create a microenvironment for sessile cells which is conditioned by the nature of the EPS matrix. In general, the proportion of EPS in biofilms can vary between 50 and 90 % of the total organic matter (Christensen and Characklis 1990; Nielsen et al. 1997).

By definition, EPS are located at or outside the cell surface independent of their origin. The extracellular localization of EPS and their composition may be the result of different processes; such as, active secretion, shedding of cell surface material, cell lysis, and adsorption from the environment. Microbial EPS are biosynthetic polymers which consist mainly of polysaccharides and proteins, but can also contain substantial amounts of DNA, lipids, glycolipids and humic substances (Jahn and Nielsen 1995; Nielsen et al. 1997). Most bacteria are able to produce EPS, whether they grow in suspension or in biofilms. Cell surface polymers and EPS are of major importance for the development and structural integrity of flocs and biofilms. They mediate interactions between the cells and maintain the three-dimensional arrangement. It must be pointed out that polysaccharides are not necessarily the main EPS component. However, not much is known about synergistic gelling of polysaccharides, proteins and humic substances. In many cases of environmental biofilm samples, proteins prevail, and humic substances are also integrated in the EPS matrix, being considered by some authors as belonging to the EPS (Wingender et al. 1999a).

Much information has been collected about the chemical nature and physico-chemical properties of extracellular polysaccharides, since they are abundant in bacterial EPS. Specific polysaccharides (e.g. xanthan) are only produced by individual strains, whereas non-specific polysaccharides (e.g. levan, dextran or alginate) are found in a variety of bacterial strains or species (Christensen and Characklis 1990). Non-carbohydrate moieties like acetyl, pyruvyl and succinyl substituents can greatly alter the physical properties of extracellular polysaccharides and the way in which the polymers interact with one another, with other polysaccharides or proteins, and with inorganic ions (Sutherland 1984). The network of microbial polysaccharides displays a relatively high water-binding capacity and is mainly responsible for acquisition and retention of water to form a highly hydrated environment within flocs and biofilms (Chamberlain 1997).

As described above, extracellular polysaccharides are believed to have the main structural function within biofilms by forming and stabilizing the biofilm matrix. The role of proteins, however, is mostly considered in terms of their enzymatic activity. Only a few authors speculate that extracellular proteins may also have structural functions. For example, the bridging of extracellular polysaccharides by lectin-like proteins is discussed (Dignac et al. 1998). Furthermore, the role of lectins (proteins with specific binding-sites for carbohydrates) in adhesion of bacterial cells and biofilm formation has been investigated (Tielker et al. 2005). A portion of the extracellular proteins has been identified as enzymes. An overview of extracellular enzymes can be found in Wingender and Jaeger (2002). Enzyme activities in biofilms include among others aminopeptidases, glycosidases, esterases, lipases, phosphatases and oxidoreductases (Lemmer et al. 1994; Frølund et al. 1995). Most of these enzymes are an integrated part of the EPS matrix (Frølund et al. 1995). It is believed that their main function is the extracellular degradation of macromolecules into low molecular weight compounds, which can then be transported into the cells and are available for microbial metabolism. The EPS matrix prevents loss of the enzymes and the degradation products and keeps them in close proximity to the biofilm cells. Moreover, specific interactions between extracellular enzymes and other EPS components have been observed resulting in the protection and localization of the enzyme (Wingender 1990; Wingender et al. 1999b). It is suggested that the structure of the EPS matrix might not be purely random but is involved in the regulation and activity of extracellular enzymes. Thus, the cell maintains a certain level of control over enzymes which are otherwise out of reach.

6.2.3 Environmental Role of Biofilms

Biofilms are involved in the geochemical cycles of most elements (Ehrlich 2002). They contribute strongly to the primary production of biomass on earth, and in particular to microbial mats (Stal and Caumette 1994). They also represent the "global cleaning machine" for all biological material which is mineralized. As most of the biomass is not dissolved but particulate, their utilization as substrate has to be performed by attached microorganisms, i.e., biofilms, and their extracellular enzymes. Thus, biological degradation is chiefly the result of the microbial activity in biofilms.

According to Whitman et al. (1998), approx. 500 billion tons of organic carbon are bound in prokaryotes, which represent about half the carbon present in biomass. The number of prokaryotes is estimated at $4 - 6 \times 10^{30}$ cells, most of which (90 – 95 %) occur in sediments, and thus in biofilms.

Soil microorganisms are an important component of the soil that performs many processes associated with energy transfer and nutrient cycling (Horwath 2002). These functions are critical to maintaining ecosystem productivity at all levels of the food web. The soil microbial biomass takes advantage of the multitude of soil niches providing different habitats and substrates. The most important function of soil biofilms is the decomposition of organic material, thus generating nutrients for their own utilization or for other organisms or plants. In this respect, the soil biofilms act as a source and sink for nutrients in the soil. Another important by-product beyond nutrient cycling is the formation of stable organic matter. The soil organic matter, through its interaction with minerals, serves many functions that increase soil quality through enhancement of physical, chemical, and biological characteristics of the soil matrix.

Biofilms play an essential role in the formation of soils as well as of subsoil formations. The biogeochemical alteration of minerals affects groundwater flow and thus transport of bacteria, particles, and of soluble compounds in the subsurface (Bachofen et al. 1998). For example, soil biofilms are involved in geochemical processes, such as diagenesis, weathering, and precipitation, as well as in oxidation/reduction reactions of metals, carbon, nitrogen, sulphur and other elements (Bachofen et al. 1998). Mineral deposition is also performed by microorganisms in biofilms; especially impressive are the formations of calcium carbonate which are attributed mainly to microbial activity (Beveridge et al. 2002; Ehrlich 2002). Early work on microorganisms in rock formation has been published only recently (Feldmann 1997). Biofilms are involved in the mobilization of minerals on a geological scale, and thus provide metal ions for the biosphere. For example, acids produced by biofilm organisms will slowly dissolve carbonates within sandstone and thus enhance the weathering process. This not only decomposes the mineral but also mobilizes metal ions. The role of microorganisms on carstification has recently been reviewed by Bennett and Engel (2005).

6.3 Biofilms and their Influence on Hydraulic Conditions in Porous Media

The type of biofilm (patchy or confluent) will have an impact not only on the biodegradation of pollutants and on the movement of colloids but also on porous media hydrodynamics. An example of microbial growth within the soil matrix is shown in Fig. 6.1. A great deal of the physical characteristics of porous media biofilms will be determined by the types of microorganisms composing the biofilm, their ecology and the geochemistry of the site (Anderson and Lovley 1997).

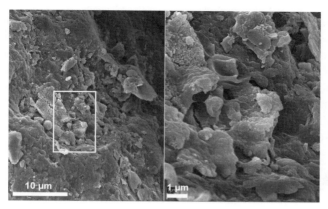

Fig. 6.1. Scanning electron micrograph of a soil matrix grain with microbial aggregates and attached inorganic colloids (courtesy of M. Strathmann)

Most pristine subsurface environments are oligotrophic. This oligotrophic nature derives from the fact that there are small amounts of organic substrates available. Most easily degraded carbon sources are used by surface or close-to-surface biota before they can reach deeper layers within the surface (Anderson and Lovley 1997). Even though microbial growth rates in subsurface environments are slower than in surface soils, there is a wide diversity of microbial life in many of them (Kieft and Phelps 1997). There are also indications that the microbial community is adapted to these oligotrophic conditions. For example, when bacteria were transported by advection in high recharge zones to deeper layers of the subsurface, they displayed increased nutrient stress as compared with native communities (Balkwill et al. 1997).

On the contrary, contaminated environments display a rich variety of redox processes (Ludvigsen et al. 1998). In this type of environment, anaerobic microbial processes are of much more relevance than aerobic processes. Due to the poor oxygen solubility, any traces of oxygen entering the system through diffusion from the surface or with recharge water will be quickly removed. Organic matter will be then utilized in a succession of electron accepting processes (Fig. 6.2). Depending on this complex biogeochemistry, if the supply of substrate is sufficient to allow for thick, continuous biofilms, the average biofilm thickness on individual porous medium particles will increase, thus decreasing the porosity and the permeability of the porous medium (Sharp et al. 1999).

Cunningham et al. (1991) found that biofilm thickness acquired a quasi-stable state after about 5 days of growth under high load substrate conditions. Porous medium porosity decreased between 50 and 96 % and permeability decreased between 92 and 98 %. Vandevivere and Baveye

(1992) found saturated hydraulic conductivity reduction due to the presence of bacterial extracellular polymers on the porous medium. They tested several mucoid strains and compared the results with non-mucoid mutants (Fig. 6.3). The production of extracellular polymeric substances had no effect on either cell multiplication within or movement through the sand columns. The reduction in permeability was therefore attributed to obstruction of pores by the polymers produced. All mucoid strains in this study caused clogging near the column inlet.

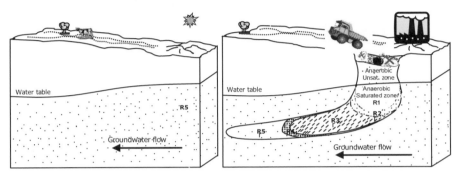

Fig. 6.2. Two extreme scenarios which can lead to the presence of patchy biofilms (left) or confluent continuous films (right). Oligotrophic conditions and O_2 as main electron acceptor on the left contrasted with high organic content and several electron accepting processes on the right caused by contamination. R1: methanogenic zone, R2: Sulfate-reducing zone, R3: Iron-reducing zone, R4: Nitrate and Manganese reducing zone, R5: aerobic zone

Polymer production decreases both the permeability and the porosity of the biofilm supporting porous medium. This fact, however, does not exclude the possibility of patchy, localized biofilms affecting the hydraulic conditions of the medium. As explained in Rittmann (1993), patchy biofilms will decrease permeability by accumulating in pore throats. Under these conditions, relatively low amounts of biomass accumulating at pore throats will cause great reductions in permeability. Preferential accumulation of biofilms at pore throats has been observed by fluorescence detection of a DNA specific dye using epifluorescence microscopy in our laboratory (Fig. 6.4).

This scenario implies that the influence of biofilms will not be the same in every situation and in every subsurface region. For the assessment of biofilm influence on colloidal transport in natural environments a deeper understanding is necessary of the biogeochemistry involved. This knowledge must be translated into appropriate parameters to be included in models that have been developed over the years for the prediction of colloidal transport in the subsurface. Porous medium heterogeneities such as biofilm formation might help to explain deviations from colloid filtration theory, CFT, which can cause some of the inconsistencies observed when predicting colloid transport (Tufenkji and Elimelech 2005).

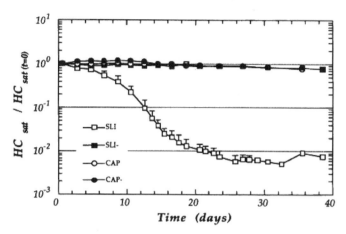

Fig. 6.3. Changes in hydraulic conductivity in columns inoculated at a 2-cm depth with a slime-producing strain (SLI), with an encapsulated strain (CAP), or with their respective non-mucoid variants (SLI⁻ and CAP⁻) (Vandevivere and Baveye 1992, Copyright with permission from American Society for Microbiology)

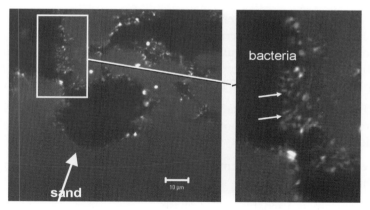

Fig. 6.4. Patchy biofilm on the surface of sand grains. The cells were stained with the DNA specific fluorochrom SYTO 9 (Molecular Probes). Epifluorescence microscopic image; bar = 10 μm

6.4 Biofilms and their Influence on Sorption Processes

A key feature of the physico-chemical properties of biofilms is their sorption behaviour (Flemming 1995; Flemming and Leis 2002; Flemming et al. 2005). Biofilms play a role both as sinks and sources of pollutants. They can adsorb water, inorganic and organic solutes, colloids and other particles. EPS, cell walls and membranes, and the cytoplasm can serve as sorption sites. These sites display different sorption preferences for certain substances, different sorption capacities, and different binding properties (Flemming et al. 1996).

Dissolved and particulate matter in water is continuously interacting at the interfaces with other phases in terms of sorption and desorption. If a biofilm is present, it will participate in such processes either by accumulation or by biochemical transformation of substances. Basically, the ability to sequester matter from the water phase is the key mechanism of nutrient acquisition for biofilm organisms. But even if substances trapped in the EPS network do not interact with the biofilm at all, they will have to pass through the biofilm when interacting with the underlying surfaces. In the case of large molecules, they can experience significant diffusion resistance (Characklis et al. 1990). The term "sorption" refers to adsorption, absorption and desorption. Adsorption implies the retention of a solute on the surface of the particles of a material. Absorption in contrast involves the retention of a solute within the interstitial molecular pores of such particles (Skoog 1996). Biofilms are involved in all of them.

The largest component in terms of volume in most biofilms is the highly hydrated extracellular polymeric matrix, understood as the network of EPS (Wingender et al. 1999a) plus the accumulated material (Flemming and Leis 2002). A closer look at biofilms in terms of sorption processes reveals a complex system (Fig. 6.5). The following sorption sites can be identified:

(i)   EPS, consisting of polysaccharides, proteins, nucleic acids, and lipids, providing charged and non-charged areas, as well as hydrophobic regions of proteins and polysaccharides

(ii) Cell walls and lipid membranes, again providing charged and non-charged areas

(iii) Cytoplasm as a separate water phase

It is obvious that each of these sites has different sorption mechanisms and capacities (Beveridge 1989). Sorption characteristics in bacteria as living organisms can change depending on a great number of factors (Langley and Beveridge 1999). For example, the extent of sorption of some heavy metals will depend on nutritional factors. Nickel uptake in

P. aeruginosa can be increased or lowered depending on the supplied carbon source (Sar et al. 1998). This is also exemplified on studies in which a biofilm of *P. putida* was exposed to toluene, which resulted in an increase of charged groups in the EPS providing more ionic binding sites (Schmitt et al. 1995). Interactions of metals and other substances with microbial cells are important for colloid and colloid-facilitated transport. A review on sorption properties of biofilms can be found in Flemming et al. (2005).

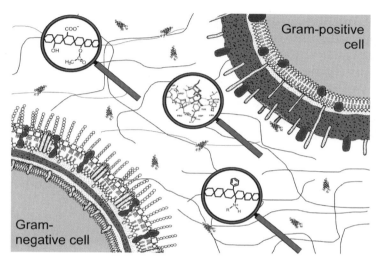

Fig. 6.5. Schematic depiction of the binding sites in biofilms (Gram-positive and Gram-negative cell, and EPS including polysaccharides and proteins)

6.5 Biofilms and their Influence on Colloid Mobility

6.5.1 Colloids in the Subsurface - from Anorganic to Biotic

The presence of mobile colloids in the subsurface has been acknowledged already since several decades (reviewed in McCarthy and McKay 2004). The focus of the studies on these type of particles has been diverse, ranging from early studies on the control of microbial contamination to the migration of clay from the surface to deeper layers in soil diagenesis (McCarthy and McKay 2004). In the field, particles such as clay particles, iron oxyhydroxides, silica, natural organic matter, viruses and bacteria have been found as mobile colloidal phases (Kretzschmar et al. 1997). The problem of deep bed filtration in wastewater treatment and other chemical and engineering tasks encouraged most of the early research on colloid transport. Data obtained from well defined collectors and colloidal particles in model systems allowed for the development of what is now known as colloid filtration theory (CFT) (Yao et al. 1971).

The conditions under which the set of equations composing the CFT approximately describes the problem of colloid transport are quite limited i.e. the porous matrix is assumed to be initially free of colloidal particles (clean bed), there must be steady state, saturated flow conditions, there should be no ripening or blocking effects and there should not be preferential flow paths. Besides, the colloid retention regime should be that of "fast" retention conditions, i.e. at moderate to high ionic strengths. These conditions are met in well controlled laboratory column studies but not in natural environments.

When these conditions are met, the advection-dispersion equation for solute transport can be adapted to colloid transport by including a term for particle retention. The concentration of suspended particles at a determined column depth and time, $C(x,t)$ can then be written as:

$$\frac{\partial C}{\partial t} = D \frac{\partial^2 C}{\partial x^2} - v \frac{\partial C}{\partial x} - kC, \qquad (6.1)$$

where v is interstitial colloid particle velocity, D is the hydrodynamic dispersion coefficient, and k is the particle deposition rate coefficient. Solutions of this equation have been used for modelling colloidal transport including microorganisms in many studies (Gross et al. 1995; Johnson et al. 1996; Kretzschmar et al. 1997; Grolimund et al. 1998; Li and Logan 1999; Ryan et al. 1999; Bhattacharjee et al. 2002; Dong et al. 2002; Saiers and Lenhart 2003). The influence of factors ranging from porous media particle charge to the influence of surfactants has been studied actively (Table 6.1). Even though CFT has been used extensively for describing microbial transport, in most realistic conditions it might not be an appropriate model to use (For an in-depth discussion on CFT and its disadvantages see Chap. 5).

Interest on colloidal movement was also catapulted by the realization that they can act as mobile phases for the transport of strongly sorbing contaminants (Ryan and Elimelech 1996). It was clear that a contaminant can then be found in three regions: dissolved, sorbed to mobile colloids and sorbed to immobile surfaces (Fetter 1999). Certain conditions have been identified which must be met for the colloid transport to be important in terms of pollutant mobility. First of all colloids must be transported through the porous medium, there must be a strong association with the contaminant and they should travel sufficiently long distances (e.g. reaching drinking water abstractions).

The study of colloid movement and its influence on the fate of common subsurface contaminants is also a microbiological endeavour. This is true not only because many microorganisms fall into the size range of colloidal particles, but mainly because their active metabolisms will have an impact on the chemical and hydrodynamic characteristics of their habitats.

Table 6.1. Colloid transport studies and their focus

Factor studied	Type of colloid	References
Porous media charge or charge heterogeneity	Latex particles, silica microspheres	(Elimelech et al. 2000; Chen et al. 2001; Elimelech et al. 2003)
Surface charge and chemistry	Hematite particles, bacteria	(Kretzschmar and Sticher 1998; Rijnaarts et al. 1999; Bolster et al. 2001; Becker et al. 2004)
Ionic strength and pH	Natural colloids, silica colloids, bacteria	(Jewett et al. 1995; Grolimund et al. 2001; Bunn et al. 2002; Saiers and Lenhart 2003; Leon-Morales et al. 2004; Redman et al. 2004)
Electrolyte type and concentration	Latex particles	(Liu et al. 1995; Davis et al. 2001)
Hydraulic conditions of porous media	Silica particles, bacteria	(Fang and Logan 1999; Ren et al. 2000; Powelson and Mills 2001)
Porous media geometry and grain size	Bacteria	(Bolster et al. 2001)
Influence of retained particles	Latex particles, bacteria	(Song and Elimelech 1993; Camesano et al. 1999)
Influence of surfactants and other organics	Bacteria	(Rogers and Logan 2000; Bolster et al. 2001; Dong et al. 2002)

Microorganisms, regardless of being dead or alive, have been known to act as accumulation sites for contaminants e.g. dissolved heavy metals, in a wide variety of environments, including the subsurface (Ferris 2000). Bacteria can migrate long distances in porous media. Mobile bacteria can also be generated by detachment from biofilms. Thus microorganisms can be described as "biocolloids". Biocolloids migrating either individually or in small aggregates through porous materials have the potential for colloid facilitated transport of contaminants. Bacteria-facilitated cadmium transport for example has been observed in alluvial gravel aquifer media (Pang et al. 2005). It was found that cadmium travelled 17 to 20 times faster in the presence of bacteria than in bacteria-free controls.

Microbial deposition and colonization of porous media particles is a well observed phenomenon (Rijnaarts 1994). Accumulation of biofilms in these environments has several important consequences: the biotransformation of contaminants, changes on the porous media hydrodynamic properties and of course influence on the movement of colloids and colloid-bound contaminants (Bouwer et al. 2000).The amount and morphology of biofilms formed in porous media can be highly variable. This variability depends on the amount of biodegradable organic matter, and on electron donors and electron acceptors available.

6.5.2 Retention of Colloids in Presence of Biofilms

Influence of Ionic Strength on the Mobility of Colloids and the Stability of Biofilms in Subsurface

It is already accepted that physico-chemical conditions have great influence on the movement of colloids and from these the ionic strength and the type of cations composing the solution will excel as having a profound effect on the mobility of colloidal particles (Jewett et al. 1995; Liu et al. 1995; Grolimund et al. 1996; Li and Logan 1999; Bunn et al. 2002; Saiers and Lenhart 2003; Leon-Morales et al. 2004). Many of these studies have found that in general high salt concentrations will promote deposition of colloidal particles while low salt concentrations will promote transport. This can be explained by changes in the size of the particles' electrical double layer. The type of dominant cations, for instance, mono- or divalent cations have an important effect on the remobilization of colloids.

Two types of deposition mechanisms can be observed depending on the size of colloid aggregates formed: (i) deposition by straining, in which the agglomerates are too big to pass through the medium pores, and (ii) deposition by interception, in which small aggregates or even individual colloids are retained on the surface of collectors (e.g. sand grains) after collision. Interception mechanisms are dominated by physico-chemical forces.

Leon-Morales et al. (2004) investigated the mobility of the model colloid laponite RD in sand columns under different ionic strengths. The breakthrough curves (BTCs) recorded at the column outlet and the collision efficiencies calculated from these BTCs are shown in Fig. 6.6 (left) and Fig. 6.6 (right), respectively. Collision efficiency, α, refers to the probability of a particle to attach upon collision with a substratum. Collision efficiency was calculated as the ratio between the obtained deposition rate constant and the deposition rate constant at high salt levels (favourable deposition conditions). The deposition rate constants were obtained from the integrated amount of particles at the column effluent, the amount of particles in the influent and the average travel time of the particles through the sand column, as described in (Grolimund et al. 2001). It was found that the mobility of laponite RD decreases with increasing ionic strength, which can be explained by an increasing aggregation of the colloid and thus an enhanced retention within the sand matrix. This behaviour is in accordance with the literature (Jewett et al. 1995; Saiers and Lenhart 2003).

Most results of colloid transport studies have been obtained from column experiments investigating suspended particle concentrations from influents and effluents only. The reduction in suspended particle concentration in most cases has been believed to follow CFT. Studying retention patterns of colloids inside the matrix allowed researchers to demonstrate inconsistencies with the predictions of CFT even at well defined and controlled physico-chemical conditions (Tufenkji and Elimelech 2005).

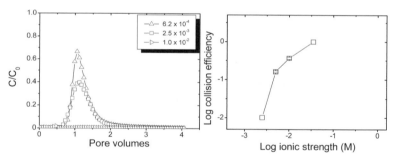

Fig. 6.6. Left: breakthrough curves of laponite RD in sterile sand columns under varied ionic strength (NaCl); Right: collision efficiencies calculated from the BTCs on the left (Leon-Morales et al. 2004, Copyright with permission from Elsevier)

To study the mobility of colloids within a porous matrix in situ Leon-Morales et al. (2004) developed a microscope flow chamber model of a sand column for Confocal Laser Scanning Microscopy, CLSM. The transport of laponite RD labelled with the fluorochrome rhodamine 6G within a sterile sand matrix was visualized at various ionic strengths. Time series of the transport of laponite in deionised water and at 10 mM NaCl are shown in Fig. 6.7. It can be seen that the injected pulse of the colloid is transported without retention through the sand matrix in a peak-like form in the case of deionised water (Fig. 6.7, upper panel). On the other hand, the colloid tends to aggregate in the case of elevated ionic strength (10 mM NaCl) and attaches to the sand surface, thus being retained within the sand matrix (Fig. 6.7, lower panel).

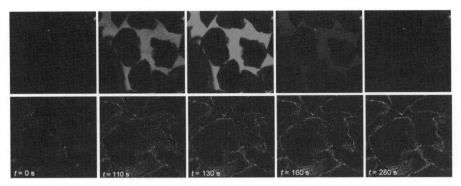

Fig. 6.7. Transport of laponite RD through a sterile sand matrix in dependence of the ionic strength; upper panel: deionised water; lower panel 10 mM NaCl. CLSM time series of a sand-filled microscopic flow chamber are shown, magnification: 200-fold, image size 460 x 460 μm

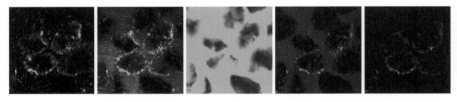

Fig. 6.8. Detachment of biofilm compartments of a *P. aeruginosa* SG81 biofilm on sand during decrease of ionic strength from 70 mM to 0 mM NaCl. CLSM time series of a sand-filled microscopic flow chamber are shown; Cell were stained with SYTO 9 (Molecular Probes); Magnification: 200-fold, image size 460 x 460 μm (modified from Leon-Morales et al. 2004)

The stability of biofilms grown within a sand matrix can be described in a similar manner and is dependant on the ionic strength as well as the predominant type of cations (mono-/divalent) (Leon-Morales et al. 2004). For example, a *P. aeruginosa* SG81-biofilm grown in a sand matrix is strongly attached to the sand matrix under high ionic strength. But if the ionic strength decreases, parts of the biofilm matrix will detach and become remobilized (Fig. 6.8). In this way the biofilm cells behave like biocolloids.

Influence of Biofilms on the Retention of Colloids

As already mentioned above, due to the "sticky" properties of the biofilm matrix, particles (e.g. colloids) tend to be retained by biofilms. To investigate the influence of the biofilm growth on the retention of colloids we performed transport experiments using sand columns as porous matrix, *P. aeruginosa* SG81 as model biofilm organism, and laponite RD as model colloid (Leon-Morales et al., in prep.). Biofilms were grown within the sand matrix for one, two, and three weeks. The increasing biofilm growth over time resulted in an increasing colloid retention as can be seen in the recorded BTCs (example in Fig. 6.9).

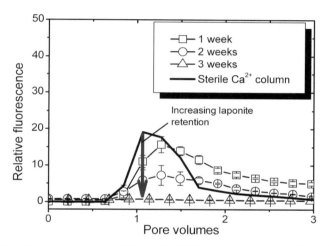

Fig. 6.9. Breakthrough curves of laponite RD in dependence of biofilm growth (Leon-Morales et al., in prep.). Biofilms of *P. aeruginosa* SG81 were grown for one, two, and three weeks in sand filled columns (Leon-Morales et al. 2004)

The transport of laponite RD in sterile and non-sterile (with biofilm growth) columns was investigated also in dependence of the present dominant cations; Na^+ and Ca^{2+} were used as monovalent and divalent cations, respectively. Collision efficiencies were calculated from the recorded BTCs as explained in Sect. "Influence of ionic strength on the mobility of colloids and the stability of biofilms in subsurface" and they are shown in Fig. 6.10 with the respective cell counts and EPS data from the columns. EPS analyses were performed according to Wingender et al. (2001). The results indicate that:

(i) the biofilm growth within the porous sand matrix increased over time,

(ii) the amount of EPS increased significantly with time,

(iii) the cell counts and the amount of EPS, especially proteins, were drastically higher in columns with biofilm grown under presence of calcium ions compared with ones grown under presence of sodium ions,

(iv) the collision efficiencies increased over time, especially in the case of experiments with calcium ions predominating.

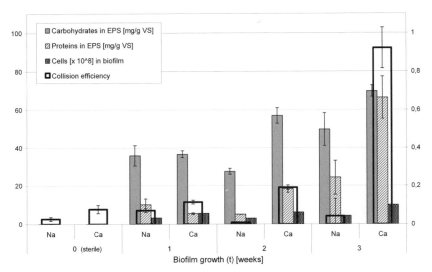

Fig. 6.10. Collision efficiencies (right *y* axis) for laponite RD in dependence of biofilm growth in correlation with extracellular polymeric substances and cell counts (left *y* axis). Biofilms of *P. aeruginosa* SG81 were grown in sand columns for one, two, and three weeks

This led to the conclusion that the increasing biofilm growth resulted in a higher retention of colloids and that this effect can be correlated with the increasing amount of proteins within the EPS. Thus, it can be speculated that the proteins act as a binding site for laponite and promote retention within the sand matrix. This would be remarkable since the "sticky" properties of a biofilm are often related only to the extracellular polysaccharides within the EPS matrix (Sutherland 2001). Further, the stability of the biofilm matrix and the interaction of laponite with the biofilm matrix are strongly enhanced by the presence of the divalent calcium ions. Under the predominance of the monovalent sodium ions, the interaction between colloids and the biofilm was limited due to destabilization, detachment and coelution of biofilm aggregates with sorbed colloids.

Literature regarding a direct quantification of porous media biofilm influence on colloidal transport is scarce; however, studies on microbial transport and biofilm influence are available. Hozalski and Dai (2001) found that the presence of Natural Organic Matter, NOM and biofilms resulted in a significant decrease in oocyst removal efficiency as compared with biofilm absent replicates. Zeta potential measurements indicated that NOM may have enhanced electrostatic repulsion between the oocyst particles and the negatively charged glass filter media. The influence of steric effects was also taken into account.

Biofilm structure has a remarkable impact on mass transfer processes. The realization that biofilm structures contain voids and channels supporting convective transport processes rather than only diffusion brings important connotations for the study of biofilm influence on colloid transport. Convective mass transfer is much faster than diffusion. This could help explain the observed rapid transport of particles through biofilms (Okabe et al. 1998) or increased particle attachment rates (van Benthum et al. 1995). In porous media this can be important, especially for confluent biofilms in which convection could represent an active transport process in systems previously conceived as "diffusion limited". (Stoodley et al. 1999).

6.5.3 Remobilization of Biofilm-Bound Colloids

Besides the role of biofilms as sorption sites and sinks for colloids and colloid-bound pollutants, they also play a role in their remobilization acting as a source for colloids and contaminants. This is of considerable importance for the mobility of contaminants and the risk assessment of biofilm-bound pollutants e. g. in soils, sediments, and aquifers.

As said before, few transport studies on the fate of colloidal contaminants and colloid-bound pollutants in the subsurface take into account the presence of biofilms (Hozalski and Dai 2001). It is generally accepted that biofilms act as a sorption site for several contaminants like organic compounds and heavy metals (Flemming et al. 1996; Sar et al. 1998; Wolfaardt et al. 1998; Flemming and Leis 2002), but the role of biofilms on the sorption and remobilization of colloids has been specifically investigated only recently (Hozalski and Dai 2001; Leon-Morales et al. 2004).

Behaviour of Model Colloids in Laboratory Columns

To elucidate the role of environmental changes in ionic strength and ionic composition on the remobilization of previously bound colloids in porous media we investigated the remobilization of the model colloid laponite RD from sand columns during the decrease of ionic strength, both, under predominance of sodium and calcium ions. The sand columns were loaded with the colloid under high ionic strength conditions, thus causing retention of the colloid within the matrix. Subsequently, the ionic strength was decreased stepwise and during each step the elution profile of laponite was recorded in the column effluent online.

It was observed that the colloid was easily remobilized under the predominance of the monovalent sodium ions already during the first step of ionic strength decrease from 70 mM to 10 mM NaCl (Fig. 6.11, left). In contrast, no remobilization of the colloid occurred during the first steps of ionic strength decrease under the predominance of divalent calcium ions (Fig. 6.11, right). Only the change of sodium concentration from 0.6 mM to deionised water caused a remobilization of the colloid.

This led to the conclusion that divalent cations promote the aggregation of colloids and the interaction with the soil matrix, thus causing stronger retention even under changing ionic strength conditions. Further, it could be concluded that a change of e.g. a calcium dominated environment to sodium predomination, for example during seawater infiltration or thawing salt drainage, can cause a significant remobilization of previously retained colloids, even without lowering the ionic strength.

Behaviour of Natural Colloids

Since the soil matrix including the mineral solid phase, mobile or immobilized colloids, biofilms and soluble substances is a very complex system, the remobilization behaviour of colloids in natural environments may differ from simplified model laboratory studies.

The remobilization of natural colloids and microorganisms from a natural system was observed under the same conditions as those described in Sect. "Behaviour of model colloids in laboratory columns". For this study drilling cores from the sand matrix of a slow sand filtration site including the "Schmutzdecke" from a drinking water purification plant were used, including their native mixed biofilm population and natural colloids. After an equilibration of the system with either NaCl or $CaCl_2$ containing buffer, the occurrence of colloids and microorganisms during ionic strength decreases was monitored in the effluent.

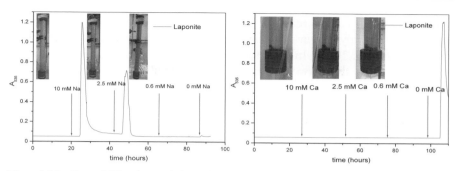

Fig. 6.11. Remobilization of laponite during the stepwise decrease of ionic strength. Left panel: NaCl, right panel: CaCl$_2$

Also under these conditions, a high remobilization of colloids as well as of biofilm compartments was observed under sodium predomination (Fig 6.12, left). As already observed in the model system (see Sect. "Behaviour of model colloids in laboratory columns") most of the remobilization occurred during the first steps of ionic strength decrease. Different types of microorganisms and colloidal aggregates were observed microscopically in the effluent during each step of ionic strength decrease (data not shown).

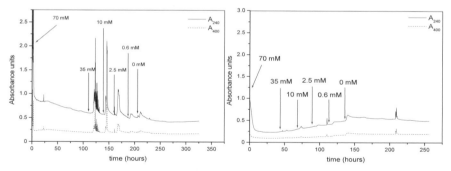

Fig. 6.12. Remobilization of colloids and biofilm components from a slow sand filter sample during the stepwise decrease of ionic strength. Left: NaCl, Right: CaCl$_2$.

In agreement with the laboratory model experiments only a very low remobilization was observed under the dominance of calcium ions even down to deionised water ionic strength levels (Fig. 6.12, right). These results confirm that divalent cations, like calcium, enhance the stability of biofilms and increase the sorption and retention of colloids within the biofilm matrix. The increase of the biofilm stability by divalent calcium ions has already been demonstrated by rheological measurements (Körstgens et al. 2001). In this study it was found that the concentration of calcium ions in the growth medium strongly influenced the mechanical properties of biofilms of *P. aeruginosa* grown on agar plates, if a certain critical calcium concentration was exceeded.

6.6 Conclusions

During e.g. rain water seepage or infiltration of road runoff contaminants can not only occur in soluble form but can also be infiltrated in colloidal form or bound to colloids. The transport of such colloids in natural environments will take place under highly varying conditions and is influenced by various factors and interactions. For example variations in ionic strength, presence of certain ions, and the presence of biofilms have to be taken into account.

The results shown in Sect. 6.5.3 indicate that these parameters have a significant influence on the sorption and retention of colloids as well as on their remobilization. It was confirmed that the presence of biofilms causes an increased colloid retention. The degree of this retention is dependent on the ionic strength and the type of cations present (mono-/divalent). Furthermore, the increased colloid retention in biofilm grown systems can be related to the increasing amount of EPS, in particular of proteins present in the porous matrix.

The results as presented above can be considered under different conditions with regard to the interactions between colloids and biofilms within the subsurface (Fig. 6.13):

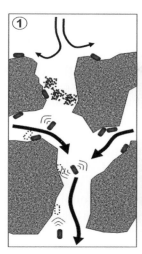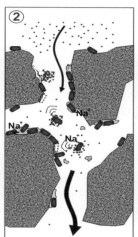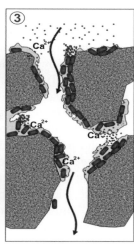

Fig. 6.13. Schematic depiction of three model transport mechanisms: (1) Indirect hydrodynamic influences: clogging and remobilization (colloid generation); (2) Direct interactions: sorption and coelution; (3) Direct interactions: sorption and deposition

- At **high ionic strength conditions** colloid movement is highly dependent on fast deposition kinetics and colloids would be quickly filtered out by interception or straining mechanisms. The presence of biofilms under these conditions can influence colloidal movement by stimulating colloid generation. If colloids are infiltrated into the subsurface under high ionic strength conditions, the colloids tend to aggregate and thus can cause clogging of pore spaces (straining). This will increase the flow velocity in the remaining pores which can cause detachment of adherent cells or previously adsorbed contaminants and/or colloids (Fig. 6.13, left panel).
- Under **low ionic strength conditions**, depending upon the type of cations predominating prior to the establishment of low ionic strength, the following interactions between colloids and biofilms can occur:
- Direct interactions - sorption and coelution:
  If colloids are infiltrated into the subsurface after the **predominance of monovalent cations**, the biofilm matrix will be destabilized resulting in a coelution of colloids together with detached biofilm components (Fig. 6.13, middle panel).
- Direct interactions – sorption and deposition:

If colloids are infiltrated into the subsurface after the **predominance of divalent cations,** the biofilm matrix will not be destabilized as is the case with monovalent cations. This results in a strong retention probably by sorption of colloids to the biofilm matrix. (Fig. 6.13, right panel).

From these scenarios it can be concluded that under high ionic strength conditions, retention of colloids within the subsurface is generally favoured. Here, biofilms act as sorption sites and sinks for colloids and colloid-bound pollutants. However, if a decrease in ionic strength under predominance of monovalent cations or an exchange of divalent cations to monovalent cations occurs, a remobilization of colloids can be expected. In this case, biofilms represent a source for colloidal or colloid-bound contaminants. Thus, environmental events which led to significant changes in ionic strength and/or ionic composition within the subsurface will certainly change the mobility of colloids, and thus potentially also of contaminants. For example, this will be the case during heavy rainfalls, stormwaters, seawater infiltration, tidal processes at estuaries and infiltration of thawing salts from road runoff. The infiltration of thawing salt will change the natural ionic composition in the subsurface to a sodium-dominated system. After the thawing period the ionic strength will decrease and thus will cause a remobilization of previously retained colloids and colloid-bound pollutants from road runoff, e. g. heavy metals and aromatic hydrocarbons.

Colloid and colloid-bound contaminant transport studies should take into account the metabolic activities of microorganisms living in subsurface environments and their influence on physico-chemical parameters, like e.g. the redox conditions. For example, plutonium mobilization in contaminated soil has been shown to be enhanced due to fermentative microbial activity, which causes dissolution of iron phases and consequent stabilization of Pu in colloidal form (Gillow 2004).

As depicted in Fig. 6.14, several factors of influence have impact on colloid transport. Most studies in the literature only take into account a portion of these factors. Most often only the physico-chemical and geochemical factors have been taken into account. Porous medium heterogeneities, such as biofilm formation might explain deviations from CFT, which can cause some of the inconsistencies observed when predicting colloidal transport (Tufenkji and Elimelech 2005).

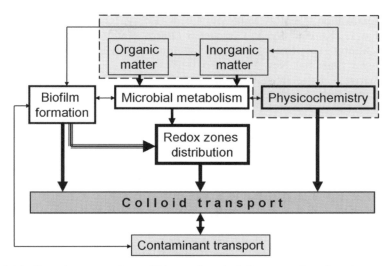

Fig. 6.14. Complete overview of biogeochemical factors to be taken into account when making predictions on colloid transport in natural subsurface environments. Shaded area: covered by conventional studies

On the background of this overview, however, for the assessment of all influence on colloidal transport in natural environments, a deeper understanding is necessary of the biogeochemistry involved.

References

Allison D (2003) The biofilm matrix. Biofouling 19(2): 139-150

Anderson RT, Lovley DR (1997) Ecology and biogeochemistry of in situ groundwater bioremediation. In: Press JP (ed) Advances in microbial ecology vol 15. New York, pp 289-333

Bachofen R, Ferloni P, Flynn I (1998) Microorganisms in the subsurface. Microbiol Res 153 (1):1-22

Balkwill DL, Murphy EM, Fair DM, Ringelberg DB, White DC (1997) Microbial communities in high and low recharge environments: Implications for microbial transport in the vadose zone. Microb Ecol 35:156-171

Becker MW, Collins SA, Metge DW, Harvey RW, Shapiro AM (2004) Effect of cell physicochemical characteristics and motility on bacterial transport in groundwater. J. Contam. Hydrol. 69:195-213

Bennett PC, Engel AS (2005) Microbial contributions to karstification. In: Gadd GM, Semple KT, Lappin-Scott HM (eds) Micro-organisms and Earth Systems - Advances in Geomicrobiology, SGM Symposium 65. Cambridge University Press Cambridge, pp 3345-3363

Beveridge TJ (1989) Role of cellular design in bacterial metal accumulation and mineralization. Ann Rev Microbiol 43:147-171

Beveridge TJ, Korenevsky A, Glasauer S (2002) Biomineralization by bacteria. In: Bitton G (ed) Encyclopedia of Environmental Microbiology, vol 2. John Wiley & Sons, Inc, New York, pp 670-681

Bhattacharjee S, Ryan JN, Elimelech M (2002) Virus transport in physically and geochemically heterogeneous subsurface porous media. J Contam Hydrol 57:161-187

Bolster CH, Mills AL, Hornberger GM, Herman JS (2001) Effect of surface coatings, grain size, and ionic strength on the maximum attainable coverage of bacteria on sand surfaces. J Contam Hydrol 50:287-305

Bouwer E, Rijnaarts HHM, Cunningham AB, Gerlach R (2000) Biofilms in porous media. In: Bryers JD (ed) Biofilms II: Process analysis and applications. Wiley-Liss, pp 123-158

Bunn RA, Magelky RD, Ryan JN, Elimelech M (2002) Mobilization of natural colloids from an iron oxide coated sand aquifer: effect of pH and ionic strength. Environ Sci Technol 36:314-322

Camesano TA, Unice KM, Logan BE (1999) Blocking and ripening of colloids in porous media and their implications for bacterial transport. Coll Surf A - Physicochem Eng Asp 160:291-308

Chamberlain AHL (1997) Matrix polymers: the key to biofilm processes. In: Wimpenny J, Handley PS, Gilbert P, Lappin-Scott H, Jones M (eds) Biofilms: Community Interactions and Control. BioLine, Cardiff, pp 41-46

Characklis WG, Turakhia MH, Zelver N (1990) Transport and interfacial transfer phenomena. In: Characklis WG, Marshall KC (eds) Biofilms. John Wiley, New York, pp 265-340

Chen JY, Ko CH, Bhattacharjee S, Elimelech M (2001) Role of spatial distribution of porous medium surface charge heterogeneity in colloid transport. Coll Surf A - Physicochem Eng Asp 191:3-15

Christensen BE, Characklis WG (1990) Physical and chemical properties of biofilms. In: Characklis WG, Marshall KC (eds) Biofilms. John Wiley, New York, pp 93-130

Cunningham AB, Characklis WG, Abedeen F, Crawford D (1991) Influence of biofilm accumulation on porous media hydrodynamics. Environ Sci Technol 25:1305-1311

Davis CJ, Eschenazi E, Papadopoulos KD (2001) Combined effects of Ca^{2+} and humic acid on colloid transport through porous media. Coll Polym Sci 280 (1):52-58

De Jonge LW, Kjaergaard C, Moldrup P (2004) Colloids and colloid-facilitated transport of contaminants in soils. Vadose Zone J 3:321-325

Dignac M-F, Urbain V, Rybacki D, Bruchet A, Snidaro D, Scribe P (1998) Chemical description of extracellular polymers: implication on activated sludge floc structure. Wat Sci Technol 38:45-53

Dong H, Rothmel R, Onstott TC, Fuller ME, DeFlaun MF, Streger SH, Dunlap R, Fletcher M (2002) Simultaneous transport of two bacterial strains in intact cores from Oyster, Virginia: biological effects and numerical modeling. Appl Environ Microbiol 68:2120-2132

Donlan RM, Costerton JW (2002) Biofilms: survival mechanisms of clinically relevant microorganisms. Clin Microbiol Rev 15 (2):167 - 193

Dworkin M (1996) Recent advances in the social and developmental biology of the myxobacteria. Microbiol Rev 60 (1):70-102

Ehrlich HL (2002) Geomicrobiology, 4th edn. Marcel Dekker, New York

Elimelech M, Chen JY, Kuznar ZA (2003) Particle deposition onto solid surfaces with micropatterned charge heterogeneity: the "hydrodynamic bump" effect. Langmuir 19 (17):6594-6597

Elimelech M, Nagai M, Ko CH, Ryan JN (2000) Relative insignificance of mineral grain zeta potential to colloid transport in geochemically heterogeneous porous media. Environ Sci Technol 34:2143-2148

Fang Y, Logan BE (1999) Bacterial transport in gas-sparged porous medium. J Environ Eng - ASCE 125 (7):668-673

Feldmann M (1997) Geomicrobial processes in the subsurface: a tribute to Johannes Neher's work. FEMS Microbiol Rev 20:181-189

Ferris FG (2000) Microbe-metal interactions in sediments. In: Riding RE, Awramik SM (eds) Microbial sediments. Springer, Berlin, pp 121-126

Fetter CW (1999) Contaminant hydrogeology, 2nd edn. Prentice Hall, New York

Flemming H-C (1995) Sorption sites in biofilms. Wat. Sci. Technol. 32 (8):27-33

Flemming H-C, Leis A (2002) Sorption properties of biofilms. In: Bitton G (ed) Encyclopedia of Environmental Microbiology, vol 5. John Wiley & Sons, Inc., New York, pp 2958-2967

Flemming H-C, Wingender J (2002) Extracellular polymeric substances (EPS): Structural, ecological and technical aspects. In: Bitton G (ed) Encyclopedia of Environmental Microbiology, vol 4. John Wiley & Sons, Inc., New York, pp 1223-1231

Flemming H-C, Wingender J (2003a) The crucial role of extracellular polymeric substances in biofilms. In: Bishop PL, Wilderer PA (eds) Biofilms in Wastewater Treatment: An Interdisciplinary Approach. IWA Publishing, London, pp 178-210

Flemming H-C, Wingender J (2003b) Biofilms. In: Steinbüchel A (ed) Biopolymers, General Aspects and Special Applications, vol 10. Wiley-VCH, Weinheim, pp 209-245

Flemming H-C, Schmitt J, Marshall KC (1996) Sorption properties of biofilms. In: Calmano W, Förstner U (eds) Sediments and Toxic Substances: Environmental Effects and Ecotoxicity. Springer, Berlin, pp 115-157

Flemming H-C, Leis A, Strathmann M, Leon-Morales CF (2005) The matrix reloaded - an interactive milieu. In: McBain A, Allison D, Pratten J, Spratt D, Upton M, Verran J (eds) Biofilms: Persistence and Ubiquity. The Biofilm Club, Manchester, pp 67-81

Frølund B, Griebe T, Nielsen PH (1995) Enzymatic-activity in the activated-sludge floc matrix. Appl Environ Microbiol 43 (4):755-761

Geesey GG (1982) Microbial exopolymers: ecological and economic considerations. ASM News 48:9-14

Gillow J (2004) Microbial transformation of the chemical association and mobility of actinides in contaminated soil. CEMS Graduate Student Conference, Center for Environmental Molecular Science

Grolimund D, Borkovec M, Bartmettler K, Sticher H (1996) Colloid-facilitated transport of strongly sorbing contaminants in natural porous media: a laboratory column study. Environ Sci Technol 30:3118-3123

Grolimund D, Elimelech M, Borkovec M (2001) Aggregation and deposition kinetics of mobile colloidal particles in natural porous media. Coll Surf A - Physicochem Eng Asp 191:179-188

Grolimund D, Elimelech M, Borkovec M, Barmettler K, Kretzschmar R, Sticher H (1998) Transport of in situ mobilized colloidal particles in packed soil columns. Environ Sci Technol 32:3562-3569

Gross MJ, Albinger O, Jewett DG, Logan BE, Bales RC, Arnold RG (1995) Measurement of bacterial collision efficiencies in porous media. Wat Res 29 (4):1151-1158

Horwath WR (2002) Biomass: soil microbial biomass. In: Bitton G (ed) Encyclopedia of Environmental Microbiology, vol 2. John Wiley & Sons, Inc, New York, pp 663-670

Hozalski RM, Dai X (2001) Investigation of factors affecting the removal of *Cryptosporidium parvum* oocysts in porous media filters. WRC Technical Report 142:17-24

Jahn A, Nielsen PH (1995) Extraction of extracellular polymeric substances (EPS) from biofilms using a cation exchange resin. Wat Sci Technol 32 (8):157-164

Jewett DG, Hilbert TA, Logan BE, Arnold RG, Bales RC (1995) Bacterial transport in laboratory columns and filters: influence of ionic strength and pH on collision efficiency. Wat Res 29 (7):1673-1680

Johnson PR, Sun N, Elimelech M (1996) Colloid transport in geochemically heterogeneous porous media: modeling and measurements. Environ Sci Technol 30:3284-3293

Kieft TL, Phelps TJ (1997) Life in the slow lane: activities of microorganisms in the subsurface. In: Amy PS, Haldeman, DL (eds) The microbiology of the terrestial deep subsurface. Lewis publishers, New York, pp 356

Körstgens V, Flemming HC, Wingender J, Borchard W (2001) Influence of calcium ions on the mechanical properties of a model biofilm of mucoid *Pseudomonas aeruginosa*. Wat Sci Technol 43 (6):49-57

Kretzschmar R, Barmettler K, Grolimund D, Yan Y-d, Borkovec M, Sticher H (1997) Experimental determination of colloid deposition rates and collision efficiencies in natural porous media. Water Resour Res 33 (5):1129-1137

Kretzschmar R, Sticher H (1998) Colloid transport in natural porous media: influence of surface chemistry and flow velocity. Phys Chem Earth 23 (2):133-139

Langley S, Beveridge TJ (1999) Metal binding by *Pseudomonas aeruginosa* PAO1 is influenced by growth of the cells as a biofilm. Can J Microbiol 45 (7) 616-622

Lemmer H, Roth D, Schade M (1994) Population density and enzyme activities of heterotrophic bacteria in sewer biofilms and activated sludge. Wat Res 28:1341-1346

Leon-Morales CF, Leis AP, Strathmann M, Flemming H-C (2004) Interactions between laponite and microbial biofilms in porous media: implications for colloid transport and biofilm stability. Wat Res 38:3614-3626

Li Q, Logan BE (1999) Enhancing bacterial transport for bioaugmentation of aquifers using low ionic strength solutions and surfactants. Wat Res 33 (4):1090-1100

Liu D, Johnson PR, Elimelech M (1995) Colloid deposition dynamics in flow through porous media: role of electrolyte concentration. Environ Sci Technol 29:2963-2973

Ludvigsen L, Albrechtsen H-J, Heron G, Bjerg PL, Christensen TH (1998) Anaerobic microbial redox processes in a landfill leachate contaminated aquifer (Grindsted, Denmark). J Contam Hydrol 33:273-291

McCarthy JF, McKay LD (2004) Colloid transport in the subsurface: past, present, and future challenges. Vadose Zone J 3:326-337

Nielsen PH, Jahn A, Palmgren R (1997) Conceptual model for production and composition of exopolymers in biofilms. Wat Sci Technol 36 (1):11-19

Okabe S, Kuroda H, Watanabe Y (1998) Significance of biofilm structure on transport of inert particulates into biofilms. Wat Sci Technol 38 (8-9):163-170

Pang L, Close ME, Noonan MJ, Flintoft MJ, van den Brink P (2005) A Laboratory study of bacteria-facilitated cadmium transport in alluvial gravel aquifer media. J Environ Qual 34 (1):237-247

Powelson DK, Mills AL (2001) Transport of *Escherichia coli* in sand columns with constant and changing water contens. J Environ Qual 30:238-245

Redman JA, Walker SL, Elimelech M (2004) Bacterial adhesion and transport in porous media: role of the secondary energy minimum. Environ Sci Technol 38:1777-1785

Ren J, Packman AI, Welty C (2000) Correlation of colloid collision efficiency with hydraulic conductivity of silica sands. Water Resour Res 36 (9):2493-2500

Rijnaarts HHM (1994) Interactions between bacteria and solid surfaces in relation to bacterial transport in porous media. PhD thesis, Wageningen University, Wageningen, The Netherlands

Rijnaarts HHM, Norde W, Lyklerna J, Zehnder AJB (1999) DLVO and steric contributions to bacterial deposition in media of different ionic strengths. Coll Surf B - Biointerfaces 14:179-195

Rittmann BE (1993) The significance of biofilms in porous media. Water Resour. Res 29 (7):2195-2202

Rogers B, Logan BE (2000) Bacterial transport in NAPL-contaminated porous media. J Environ Eng - ASCE 126 (7):657-666

Ryan JN, Elimelech M (1996) Colloid mobilization and transport in groundwater. Coll Surf A - Physicochem Eng Asp 107:1-56

Ryan JN, Elimelech M, Ard RA, Harvey RW, Johnson PR (1999) Bacteriophage PRD1 and silica colloid transport and recovery in an iron oxide-coated sand aquifer. Environ Sci Technol 33:63-73

Saiers JE, Lenhart JJ (2003) Ionic-strength effects on colloid transport and interfacial reactions in partially saturated porous media. Water Resour Res 39 (9):1256

Sar P, Kazy SK, Asthana RK, Singh SP (1998) Nickel uptake by *Pseudomonas aeruginosa*: role of modifying factors. Curr Microbiol 37:306-311

Schmitt J, Nivens D, White DC, Flemming H-C (1995) Changes of biofilm properties in response to sorbed substances - an FTIR-ATR study. Wat Sci Technol 32 (8):149-155

Shapiro JA (1998) Thinking about bacterial populations as multicellular organisms. Annu Rev Microbiol 52:81-104

Sharp RR, Cunningham AB, Komlos J, Billmayer J (1999) Observation of thick biofilm accumulation and structure in porous media and corresponding hydrodynamic and mass transfer effects. Wat Sci Technol 39 (7):195-201

Skoog DA, West DM, Holler FJ (1996) Fundamentals of analytical chemistry, 7th edn. Saunders College Publishing, Philadelphia

Song L, Elimelech M (1993) Calculation of particle deposition rate under unfavourable particle-surface interactions. J Chem Soc Farad Trans 89 (18):3443-3452

Stal LJ, Caumette P (1994) Microbial mats; structure, development and environmental significance. Springer, Berlin

Stoodley P, Boyle JD, DeBeer D, Lappin-Scott HM (1999) Evolving perspectives of biofilm structure. Biofouling 14 (1):75-90

Sutherland IW (1984) Microbial exopolysaccharides - their role in microbial adhesion in aqueous systems. CRC Crit Rev Microbiol 10:173-201

Sutherland IW (2001) Biofilm exopolysaccharides: a strong and sticky framework. Microbiology 147 (Pt 1):3-9

Tielker D, Hacker S, Loris R, Strathmann M, Wingender J, Wilhelm S, Rosenau F, Jaeger K-E (2005) *Pseudomonas aeruginosa* lectin LecB is located in the outer membrane and is involved in biofilm formation. Microbiology 151:1313-1323

Tufenkji N, Elimelech M (2005) Breakdown of colloid filtration theory: role of the secondary energy minimum and surface charge heterogeneities. Langmuir 21:841-852

van Benthum WAJ, van Loosdrecht MCM, Tijhuis L, Heijnen JJ (1995) Solids retention time in heterotrophic and nitrifying biofilms in a biofilm airlift suspension reactor. Wat Sci Technol 32 (8):53-60

Vandevivere P, Baveye P (1992) Effect of bacterial extracellular polymers on the saturated hydraulic conductivity of sand columns. Appl Environ Microbiol 58 (5):1690-1698

Watnick P, Kolter R (2000) Biofilm, city of microbes. J Bacteriol 182 (10):2675-2679

Whitman WB, Coleman DC, Wiebe WJ (1998) Prokaryotes: the unseen majority. Proc. Natl Acad Sci USA 95:6578-6583

Wingender J (1990) Interactions of alginates with exoenzymes. In: Gacesa P, Russell NJ (eds) Pseudomonas infections and alginates. Chapman and Hall, London, New York, Tokyo, Melburne, Madras, pp 160-180

Wingender J, Jaeger K-E (2002) Extracellular enzymes in biofilms. In: Bitton G (ed) Encyclopedia of Environmental Microbiology, vol 3. John Wiley & Sons, Inc., New York, pp 1207-1223

Wingender J, Neu TR, Flemming H-C (1999a) What are bacterial extracellular polymeric substances? In: Wingender J, Neu TR, Flemming H-C (eds) Microbial extracellular polymeric substances. Springer, Berlin, pp 1-19

Wingender J, Grobe S, Fiedler S, Flemming H-C (1999b) The effect of extracellular polysaccharides on the resistance of *Pseudomonas aeruginosa* to chlorine and hydrogen peroxide. In: Keevil CW, Godfree A, Holt D, Dow CS (eds) Biofilms in the aquatic environment. Royal Society of Chemistry, Cambridge, pp 93-100

Wingender J, Strathmann M, Rode A, Leis A, Flemming H-C (2001) Isolation and biochemical characterization of extracellular polymeric substances from *Pseudomonas aeruginosa*. Methods in Enzymology 336: 302-314

Wolfaardt GM, Lawrence JR, Robarts RD, Caldwell DE (1994) The role of interactions, sessile growth, and nutrient amendments on the degradative efficiency of a microbial consortium. Can J Microbiol 40:331-340

Wolfaardt GM, Lawrence JR, Robarts RD, Caldwell DE (1998) In situ characterisation of biofilm exopolymers involved in the accumulation of chlorinated organics. Microb Ecol 35:213-223

Yao KM, Habibian MT, O'Melia CR (1971) Water and waste water filtration - Concepts and applications. Environ Sci Technol 5:1105-1112

7 Subsurface Transport of Heavy Metals Mediated by Biosolid Colloids in Waste-Amended Soils

Anastasios Karathanasis, Carey Johnson, Chris Matocha

Department of Plant and Soil Sciences, N-122K Ag. Science-North, University of Kentucky, Lexington, KY 40546, USA

Land application of agricultural and industrial solid waste products (biosolids) is being widely promoted as a cost-effective nutrient management alternative with dramatic increases worldwide in recent years (Linden, 1995). Although many biosolid materials are considered excellent sources of plant macronutrients, some contain high levels of heavy metals in labile forms that may pose a significant threat to soil and groundwater quality (McBride et al., 1997, Scancar et al, 2001). There is a general perception that many hydrophobic contaminants, such as heavy metals, readily sorb to the soil matrix and are considered to be virtually immobile, posing little danger to groundwater supplies (McCarthy and Zachara, 1989). However, the risk may be greater than anticipated considering that a significant metal load may be associated with dispersed biosolid colloid particles, which may be mobilized through soil macropores to lower soil depths and greater distances (Kretzschmar et al., 1995; Gove et al., 2001). The high organic carbon content and increased surface reactivity of mobile biosolid colloids may out-compete the soil matrix in heavy metal sorption affinity, thus facilitating metal transport (Karathanasis, 1999; Karathanasis and Ming, 2002). This colloid-mediated metal transport may explain the negative metal mass balances found by some researchers trying to account for the losses of biosolid-applied heavy metals in soils (McGrath and Lane, 1989; Baveye et al., 1999).

Organically enriched and organic-coated colloids, such as biosolid colloids, have been shown to enhance colloid stability by acting as steric stabilizers, neutralizing positive edge sites on mineral colloids (Kaplan et al., 1993). While mineral colloids tend to destabilize and flocculate more readily, organic colloids or organically coated mineral particles show considerably higher stability and mobility through electrostatic and steric repulsion forces even under high ionic strength conditions (Ryan and Gschwend, 1994). Organic mobile colloids have also been implicated as the primary vector for transporting contaminants in some subsurface environments, where they may be transported at a greater velocity than conservative tracers due to size exclusion effects (Kretzschmar et al., 1995, Karathanasis and Ming, 2002). Studies with undisturbed soil columns have demonstrated rapid elution of Cd, Zn, Cu, and Pb, in the presence of organics due to preferential flow processes (Camobreco et al., 1996). In other field and laboratory studies involving biosolid applications, soluble and colloidal organic constituents exhibited considerable ability to facilitate metal transport through soil porous media (del Castillo et al., 1993; Persicani, 1995). Li and Shuman (1997) reported enhanced mobility of Cd, Zn, and Pb associated with a poultry litter extract as a result of induced solubilization or the formation of soluble metal-organic complexes derived from the poultry litter. Lime-stabilized biosolid colloids have also been shown to significantly enhance heavy metal mobility in the soil, either by chemisorption or coprecipitation mechanisms onto the colloids or by the formation of soluble metal-organic complexes, particularly at high pH ranges (Karathanasis and Ming, 2002). Soil Cd mobility evaluations in field and laboratory experiments following biosolid applications suggested at least two-fold increases, mainly in the form of soluble organic complexes, with the main sorption pools being the organic and Fe-oxide fractions (Lamy et al., 1993; Hettiarachchi et al., 2003). These patterns emphasize the alarming consequences of colloid-enhanced metal transport and the ramifications for predicting the real risks of groundwater pollution by heavy metals associated with applied biosolid amendments.

The objectives of this study were: (i) to assess the stability and mobility of three biosolid colloid suspensions through selected undisturbed soil monoliths, (ii) to evaluate the mobility of Cu, Zn, Pb, Cd, Cr and Mo in association with the transported biosolid colloids, and (iii) to evaluate colloid and soil properties enhancing or inhibiting metal transport.

7.1 Materials and Methods

7.1.1 Biosolid Samples

Bulk samples of a lime-stabilized biosolid (LSB) material in the final processing stage were obtained from a municipal wastewater treatment plant in Winchester (Clark County), Kentucky. A second aerobically digested biosolid (ADB) material (in final processing stage) was obtained from the West Hickman wastewater treatment plant in Fayette County, Kentucky. The third biosolid material consisted of poultry manure (PMB), collected from a Perdue chicken production facility in McLean County,

Kentucky. All samples were placed in sealable plastic bags to maintain their original moisture and stored under refrigeration.

7.1.2 Colloid Generation

Water-dispersible colloids were fractionated from bulk samples of each waste by placing 50 g of sample in a 1 L centrifuge bottle and filling with de-ionized water. The slurry was mixed with a reciprocating shaker for 1 hr and centrifuged at 130 x g for 3.5 min. The process was repeated twice for each 50 g of waste. The colloid particles remaining in suspension were decanted and saved as a stock solution in acid washed Nalgene carboys. The colloid mass was determined gravimetrically by placing 100 mL aliquots of each stock solution into 100 mL tared Teflon beakers and drying at 100°C for 24 hr.

7.1.3 Soil Monolith Preparation

Undisturbed soil monoliths were collected from the upper Bt horizon of a Maury silt loam (fine, mixed, semiactive, mesic Typic Paleudalfs) and the AB horizon of a Woolper silt loam (fine, mixed, mesic Typic Argiudolls) at the Spindletop farm of the University of Kentucky. Another set of monoliths was collected from the Bw horizon of a Bruno fine sandy loam (sandy, mixed, thermic Typic Udifluvents) from Estill County, Kentucky. Representative soil subsamples from the three sites were also collected for physical, chemical and mineralogical characterization. The soils used in the study were selected to represent diverse textural, macroporosity, and organic matter (OM) conditions. The soil monoliths were prepared by excavating and trimming soil pedestals before encasing them in polyvinyl chloride (PVC) tubes of 18 cm diameter and 25 cm height. The annulus between the soil and the PVC tube was sealed with expandable polyurethane foam (Poly-U-Foam) to enhance monolith stability and prevent preferential flow along the PVC walls.

7.1.4 Physical, Chemical and Mineralogical Characterizations

Particle size distribution of biosolid colloid samples was determined by fractional centrifugation in the > 2, 1-2, 0.5-1, 0.2-0.1, and < 0.1 μm size range and gravimetric determinations at the following settings, respectively: 170 x g (1.5 min), 280 x g (3.0 min), 430 x g (6.0 min), 860 x g (8.5 min), and 860 x g (30 min). Electrical conductivity (EC) on a μS/cm basis and pH (in standard units) were measured on each colloid suspension with a Denver Instruments Model 250 pH*ISE*electrical conductivity meter (Arvada, CO). Organic carbon determination was performed with a Shimadzu TOC 5000A (Kyoto, Japan) carbon analyzer on approximately 100 mg of air-dried and finely ground colloid and soil samples passed through a 0.23 mm sieve. Crystalline Fe and Al oxides in colloid and soil samples were extracted with the dithionite-citrate-bicarbonate (DCB) method, and amorphous Fe-Al-oxides by the ammonium oxalate method (NRCS, 1996). Extractable Fe and Al were analyzed with a Shimadzu GFA-EX7 graphite furnace atomic absorption spectrophotometer (GFAAS) (Kyoto, Japan).

Soil particle size analysis was performed on air-dried samples using the sodium hexametaphosphate $(Na(PO_3)_6)$ pipette method (NRCS, 1996). Soil porosity estimates from bulk density measurements were used to calculate pore volumes based on the volume of each soil monolith (NRCS, 1996). Saturated soil hydraulic conductivity was determined via a modified constant head permeameter method (NRCS, 1996). Colloid- and soil-cation exchange properties were determined by the ammonium acetate method (NRCS, 1996). Mineralogical composition was performed by x-ray diffraction (XRD) and thermogravimetric (TG) analysis (Karathanasis and Hajek, 1982a). A set of Phillips PW 1840 diffractometer/PW 1729 x-ray generator (Mahwah, NJ) equipped with a cobalt x-ray tube (40 kV, 30 mA), and a Bragg-Bretano design goniometer were used for x-ray analysis. The scanning rate was set at $0.05°$ 2Θ per minute from $2°$ to $40°$ and the scattering slit at $0.1°$. A TA 2000 thermogravimetric analyzer interfaced with a 951 DuPont TG module at a heating rate of $20°$ C/min under N_2 atmosphere was used for TG analysis. The colloid surface area was estimated from thermogravimetric water sorption experiments (Karathanasis and Hajek, 1982b).

Adsorption isotherms were also generated to evaluate the affinity of the soils and biosolid colloids for Cu, Zn, Pb, Cd, Cr, and Mo. A 5 mL colloid suspension of 200 mg/L concentration was added to 10 mL Teflon test tubes containing 0-5 mg/L metal concentrations. Stock metal solutions were prepared from $CuCl_2$, $ZnCl_2$, $PbCl_2$, $CdCl_2$, $CrCl_3$, and Na_2MoO_4 reagents (> 99 % purity, Aldrich Chemicals, Milwaukee, WI). Samples were shaken on a reciprocating shaker for 24 hr at room temperature and centrifuged for one hour at (x 2750 g) 3500 rpm. Supernatants were collected and analyzed for Cu, Zn, Pb, Cd, Cr, and Mo via inductively coupled plasma atomic emission spectrometer (ICP-AES). Freundlich isotherms fitted on log-scale by linear regression were used to describe the experimental adsorption data. All analyses, except for the mineralogical characterizations, were performed on duplicate samples.

7.1.5 Colloid Leaching Experiments

Eight soil monoliths were used for each soil type, to provide replicated sets for the 3 biosolid colloids (200 mg/L), and one for the deionized water control. Prior to application of the biosolid colloid suspensions and deionized water solutions onto the soil monoliths, the biosolid colloids were spiked with 5 mg/L of Cu, Zn, Pb, Cd, Cr, and Mo and equilibrated overnight in order to increase metal loads above inherent levels and enhance metal detectability. The soil monoliths were immersed (saturated) in deionized water for approximately 24 hr to remove air pockets and loose material and then leached with 5 L of a 1 mM $CaCl_2$ solution as a conservative tracer before the application of the biosolid colloid suspensions. Leaching was conducted via unsaturated gravity flow at a rate of ~0.7 cm/hr, using a peristaltic pump. Water and suspension loads were applied on the top of each soil monolith via a sprinkler head. Natural lower boundary conditions in the bottom of the soil monoliths were simulated by suction in the elution funnel applied with a vacuum pump. The lower boundary was controlled with a Marriott device similar to that of Ritter et al. (2004), utilizing a constant level reservoir at - 10 cm suction, which was monitored with a tensiometer. Eluents were collected at 1 L intervals for a total of 21 L, corresponding to 16-24 pore volumes. Each sample was analyzed for EC and pH using a Denver Instruments Model 250 pH-ISE-conductivity meter. Chloride concentrations were measured using an adaptation of the automated ferricyanide method (APHA, 1998) on 60 µL samples. Colloid concentrations were determined turbidimetrically on each 1 L sample, using a Bio-Tek Instruments microplate reader (Bio-Tek Instruments, Winooski, VT) scanning at 540 nm. Metals associated with the eluted colloids were fractionated into soluble and sorbed forms. The soluble metal fraction was determined by filtering 25 mL of a colloid subsample through a 0.2 µm filter and saving an aliquot for metal analysis via ICP-AES. An acid extraction with 25 mL of 1 M $HCl-HNO_3$ was used to extract the sorbed metal associated with the colloids retained on the filter. Breakthrough curves of colloid and metal (total and soluble) loads eluted from each monolith were plotted as reduced concentrations (C/C_0) by pore volume for each soil.

7.1.6 Statistical Analysis

Duncan's multiple range test was used to compare differences in total metal elution among colloids and controls for different soil types. Single correlation and multiple stepwise regression analyses were conducted to establish relationships between soil and colloid properties and colloid-mediated metal transport. All statistical analyses were performed with a Statgraphics Plus version 5.0 software program (Manugistics, Inc., Rockville, MD) at $P < 0.05$ (*) and $P < 0.01$ (**) probability levels.

7.2 Results and Discussion

7.2.1 Colloid Elution

Colloid breakthrough curves (BTC) through the three soils were generally irregular, exhibiting several maxima and minima, depending on colloid and soil type (Fig. 7.1a). The irregular colloid breakthrough is attributed to cluster-type rather than uniform transport and deposition as a result of macropore blockage followed by flushing due to water pressure buildup in the blocked pores (Jacobsen et al., 1997). Overall, the colloids showed a slower breakthrough than the conservative Cl⁻ tracer, indicating considerable physical and chemical interaction in their flow path through the soil monoliths. This interaction is realized through attachment-detachment processes with the soil matrix, as well as sieving and physical entrapment of the colloids by the soil matrix. Evidence for preferential flow through soil macropores was provided by: (i) the early initial breakthrough of the conservative tracer before one pore volume was passed through the soil monoliths, and (ii) the nearly 4-5 pore volumes needed to reach $C/C_o = 1$, suggesting considerable undersaturation of the soil matrix with the tracer during the first few pore volumes of leaching (Bouma, 1991).

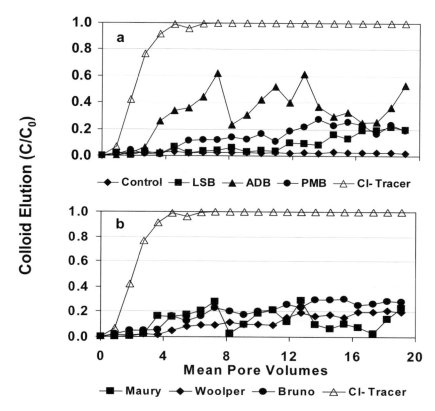

Fig. 7.1. Breakthrough curves (BTC's) for a Cl⁻ tracer and colloid elution in the absence of colloids (control) and in the presence of LSB, PMB, and ADB biosolid colloids for all soils combined (a) and for all biosolid colloids combined through the Maury, Woolper, and Bruno soils (b).

Although individual colloid recoveries through the soils varied from 0-100 % at different stages of leaching, average colloid elution (all colloids combined) among soils was much less variable (C/C_o = 0.15-0.25), with the higher elution range representing the Bruno soil and the lower the Maury and Woolper soils (Fig. 7.1b). However, when considering all soils together, the ADB colloids showed a significantly higher mobility, with an average mass recovery of about 30 %, compared to 15 % for the PMB and 10 % for the LSB colloids (Fig 7.2). This indicates that differences in colloid properties overshadowed soil variability in the elution process. The ADB colloids proved to be particularly mobile through the Maury and Bruno soils, with recoveries of 38 % and 54 %, respectively (Fig. 7.2.). These colloids showed a rapid but erratic breakthrough after 3 pore volumes, reaching maxima at 7, 11, 13, and 19 pore volumes (Fig. 7.1.). This is indicative of preferential flow patterns with alternating macropore clogging and flushing cycles dominated by colloid straining and detachment events (Jacobsen et al., 1997). The erratic elution pattern through the Maury soil was associated with pH (6.5 to 4.0) and EC (6 to 21μS/cm) fluctuations of the colloid suspensions. The mobility of the ADB colloids through the Bruno soil may have been enhanced by the high negative surface charge of the colloids and the relatively high soil OM content (Table 7.1), which caused increased electrostatic repulsion with the soil matrix. Contrary to the Maury and Bruno soils, the recovery of ADB colloids through the Woolper soil was limited to < 10 %, in spite of uniform elution chemistry. Apparently, physical straining processes overshadowed the influence of chemical factors in this experiment.

In contrast, the Woolper soil showed the highest recoveries for the PMB colloids (~ 27 %), compared to < 10 % for ADB and LSB (Fig. 7.2). PMB colloid elution was much more gradual than that of the ADB colloids with moderate fluctuations after the 10th pore volume reaching maxima of 0.2 C/C_o (Fig. 7.1). The latter coincided with sizeable pH fluctuations in eluent suspensions (8.0 to 5.0) that enhanced or retarded colloid mobility. The low pH and hydraulic conductivity of the Maury and Bruno soils may be responsible for inhibiting the mobility of PMB colloids and limiting their recoveries to < 10 % (Table 7.1 and Fig. 7.2).

The mobility of LSB colloids was the lowest among the biosolid colloids studied with total recoveries below 10 % and less erratic BTC's (Fig. 7.1). The relative immobility of the LSB compared to other colloids may be due to carbonate dissolution during the leaching cycle, as evidenced by the pH drop of the eluted suspensions from 11.3 to < 7.0 and increase in the EC (35-52 μS/cm). It is likely that the increased levels of the released Ca in solution induced colloid coagulation and inhibited their mobility (Degueldre et al., 1996).

It has to be kept in mind that most of the elution experiments can only lead to operationally defined results due to the complexity of the investigated systems. Even though this includes a limited comparability model, calculations based on thermodynamic equilibrium can supply valuable interpretations for experimental data. Including the kinetics of adsorption/desorption processes will complicate the description of the interactive system and will be the challenge for future work.

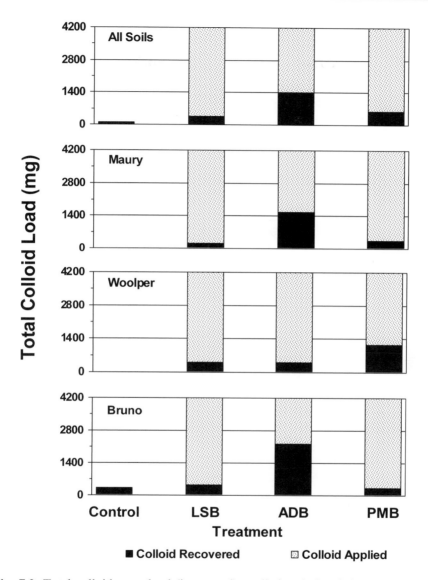

Fig. 7.2. Total colloid mass load (integrated) applied and eluted through all soils combined and each individual soil for different colloids.

Table 7.1. Selected physicochemical and mineralogical properties of colloids and soils used in the study.[a]

Properties	Soils			Colloids		
	Maury	Woolper	Bruno	Lime-stabilized biosolid	Aerobically digested biosolid	Poultry manure biosolid
Particle size distribution[b] (%)	$S_6 Si_{70} C_{24}$	$S_{11} Si_{60} C_{29}$	$S_{65} Si_{23} C_{12}$	-	-	-
Soil Porosity	0.46	0.35	0.30	-	-	-
Saturated hydraulic conductivity (cm min^{-1})	3.0	3.3	2.5	-	-	-
pH	5.6	6.1	4.0	11.3	7.2	5.5
Cation exchange capacity (CEC) (cmol kg^{-1})	17.9	24.3	10.5	80.4	101.0	93.2
S.A.R.[b]	0.21	0.20	0.45	0.17	2.92	2.34
Electrical Conductivity (EC) (μS cm^{-1})	0.2	0.4	0.3	68.2	16.3	12.5
Fe-oxides (crystalline+amorphous) (g kg^{-1})	27.9	28.9	11.2	1.3	20.1	1.1
Al oxides (crystalline+amorphous) (g kg^{-1})	7.7	7.8	1.6	4.1	11.5	1.2
Surface area (m^2 g^{-1})	93	136	38	361	506	578
Mean colloid diameter (μm)	---	---	---	0.41	0.46	0.38
O.M.[b] (%)	0.6	3.9	3.0	20.1	60.6	47.9
Mineralogical composition[b] (%)	$MV_{22}K_6Q_{65}$	$MV_{28}K_5Q_{55}$	$MV_{30}K_5Q_{60}$	$MV_{10}CC_{50}Ca_{10}$	$MV_{15}K_3CC_5$	$MV_{15}K_{15}Q_5$
K_f (Cd)[b] (L kg^{-1})	3.9	7.4	4.5	20.2	1.5	5.4
K_f (Zn)[b] (L kg^{-1})	5.4	9.2	4.8	15.6	1.3	1.8
K_f (Pb)[b] (L kg^{-1})	2.0	3.1	2.1	15.5	5.6	3.7
K_f (Cu)[b] (L kg^{-1})	3.2	8.2	2.9	1.5	1.5	2.1
K_f (Cr)[b] (L kg^{-1})	2.3	6.6	2.6	25.1	5.3	1.5
K_f (Mo)[b] (L kg^{-1})	0.0	0.0	0.0	0.0	0.0	0.0

[a] Particle size distribution, soil porosity, and hydraulic conductivity values represent the average of duplicate samples with a standard error of < 8%; chemical analyses, surface area, and mean colloid diameter values represent the average of duplicate samples with a standard error of < 5 %; mineralogical compositions represent analyses of single samples.
[b] S = sand; Si = silt; C = clay; S.A.R = sodium adsorption ratio; O.M. = organic matter; K_f = Freundlich coefficients; MV = mica + vermiculite; K = kaolinite; Q = quartz; CC = calcium carbonate; Ca = Ca(OH)$_2$.

7.2.2 Metal Elution in Association with Biosolid Colloids

Figs. 7.3 and 7.4 depict the mass-balance for metal forms (soluble and sorbed) recovered during the leaching experiments compared to the applied metal load for each colloid and control treatment. Average mass loads for colloid and metal elution calculated from replicated breakthrough curves (BTCs) in the presence of biosolid colloids for all soils combined are shown in Figs. 7.5-7.7. Variability between replicated BTC's ranged from 3-17 %, with an average value of about 8%. Generally, the presence of colloids enhanced the elution of all metals (from 2 to 1,000 times) compared to control treatments (absence of colloids), which showed elution of 0-2 %. The positive effect of colloids in facilitating metal transport was evidenced by the high overall correlation coefficient between metal and colloid elution ($R^2 = 0.68$ [**]). Surprisingly, even Mo transport was enhanced in the presence of some biosolid colloids, in spite of the lack of sorption affinity indicated from isotherm experiments. Soluble metal elution was also enhanced in the presence of biosolid colloids, particularly for Cd, Zn, and Cu, although the overall correlation between eluted soluble metal- and colloid-loads was not statistically significant ($R^2 = 0.26$). Neither the colloid- nor the soil-K_f values produced significant correlations ($R^2 = 0.01 – 0.39$) with either metal load retained by soil or eluted in association with the biosolid colloids, rendering this approach meaningless in predicting metal transport through the soils studied.

Cadmium Elution

Total Cd elution was highest in association with ADB colloids, averaging 50 % for all soils combined, compared to < 10 % for PMB and LSB colloids (Fig. 7.3). These elutions were associated with fairly irregular BTCs that correlated well with colloid elution patterns ($R^2 = 0.82$[**]) (Figs. 7.5-7.7). The high organic matter content, surface area, and surface charge, as well as the moderate suspension pH associated with the ADB colloids (Table 7.1) probably contributed to higher Cd mobility. These findings are consistent with those of Li and Shuman (1997), who reported enhanced Cd mobility in soils amended with biosolids due to increased metal solubilization and co-transport processes. Eluted soluble Cd was also higher in association with ADB colloids accounting for about 22 % of the total eluted metal load compared to < 3 % in association with PMB and LSB colloids (Fig. 7.3). The high elution of soluble Cd is attributed to the fairly low sorption affinity of the ADB colloids for Cd (Table 7.1) and the high OM status of colloids and soils, which induced the formation of soluble organic-Cd complexes (Li and Shuman, 1997). This result may have been accentuated by the higher molecular weight and hydrophilic nature of the organic complexes associated with the ADB colloids, which may significantly enhance Cd complexation and mobility (Neal and Sposito, 1986). These organo-metal complexes are fairly non-adsorptive within the soil matrix and may explain the sizeable soluble metal fraction mobilized in this study (McBride et al., 1997; Denaix et al., 2001). Greater attenuation by the soil matrix and lower elution of total and soluble Cd loads in the presence of LSB and PMB colloids were caused by lower overall colloid elutions in conjunction with lower OM content. In spite of higher sorption affinities, the Cd transportability of these colloids may have been inhibited

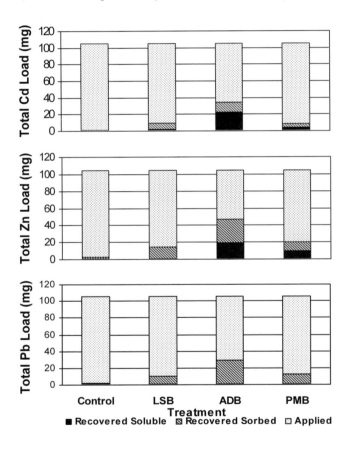

Fig. 7.3. Total mass load (integrated) of Cd, Zn, and Pb applied and recovered in soluble and sorbed form after elution through all soils combined.

by co-precipitation processes and attachment to soil matrix surfaces, particularly in association with the high pH and the carbonate surfaces present in the LSB colloids (Buchter et al., 1996). The highest amounts of Cd were transported in association with ADB colloids through the Maury and Bruno soils (33-66 %), respectively, with 27-40 % being in the soluble fraction. Only the PMB colloids through the Woolper soil and the LSB colloids through the Bruno and Woolper soils exhibited any additional meaningful elution (15-20 %). There was virtually no Cd elution in the control treatments (absence of colloids), suggesting that the colloids were the primary vectors for Cd transport through the soils studied.

Zinc Elution

Total Zn elution through all soils combined was among the highest, particularly in association with the ADB colloids, which averaged 47 %, compared to the PMB (20 %) and LSB (15 %) colloids (Fig. 7.3). Organic matter, surface area, cation exchange capacity (CEC), and carbonate colloid content in conjunction with the pH of the colloid suspensions were the most influential variables, accounting for 70 % of the variation in total Zn elution. Elution patterns were erratic, with maxima and minima correlating well with colloid BTC's (0.82^{**}) (Figs. 7.5-7.7). Although the ADB and PMB colloids in this study had the lowest overall affinity for Zn (Table 7.1), their presence induced Zn transportability, apparently due to their moderate pH and high CEC and OM content. This trend is consistent with other findings indicating that some PMB materials can greatly enhance Zn mobility in the subsurface environment (Li and Shuman, 1997). Significant amounts of soluble Zn were also eluted in the presence of ADB (18 %) and PMB (8 %) colloids attributed to their high organic matter, which enhanced soluble metal-organic complexation. Decreased Zn elution in association with LSB colloids in spite of high K_f values was caused by lower colloid elution, which was inhibited by the presence of carbonates and high ionic strength. The highest elution of Zn was mediated by the ADB colloids through the Maury and Bruno soils (45-75 %) and the PMB colloids through the Woolper soil (43 %), with nearly half of it being in the soluble fraction. The only significant Zn elution in association with the LSB colloids was observed in the Woolper and Bruno soils (15-20 %). In all other soils and control treatments with minimal or no colloid breakthrough, Zn elution was limited to < 5 %, emphasizing the important role of the colloids in the metal mobilization process.

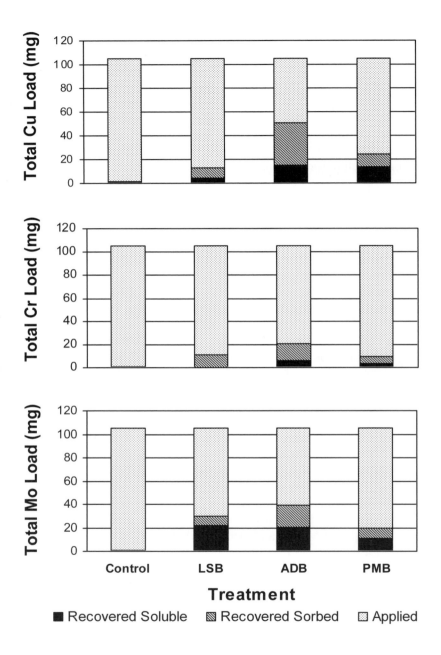

Fig. 7.4. Total mass load (integrated) of Cu, Cr, and Mo applied and recovered in soluble and sorbed form after elution through all soils combined.

Lead Elution

Quantitatively, total Pb elution through all soils combined was similar to that of Cd, with the ADB colloids eluting three times as much Pb as the PMB or LSB colloids (Fig. 7.3). Elution patterns were irregular, showing several minima and maxima corresponding to clogging and flushing episodes, which correlated well with colloid BTC's (0.85^{**}) (Figs. 7.5-7.7). The increased Pb mobility in association with the ADB colloids is consistent with high colloid organic matter content and their relatively high affinity for Pb (Table 7.1) (Amrhein et al., 1993; Denaix et al., 2001). Lower Pb affinity and colloid mobility contributed to moderate Pb transportability by PMB colloids, while for LSB colloids lower colloid elution and adverse colloid and soil properties (pH, EC, carbonates) compromised their high Pb sorption affinity (Table 7.1). In contrast with other metals that showed significant elution of soluble metal loads in the presence of colloids, > 97 % of the eluted Pb load was in the sorbed fraction. The strong affinity of the colloid surfaces for Pb is related to the low pK_a (~ 5.0) of organic colloid-OH groups, and the significant covalency of the Pb- carboxylic or Pb-phenolic groups (Logan et al., 1997; McBride, 1989). The ADB colloids through the Maury and Bruno soils and the PMB colloids through the Woolper soil were the most effective Pb carriers (32-47 %). The LSB colloids eluted through Woolper and Bruno soils transported about 12 %, while the remaining soil and control treatments showed negligible Pb mobilization.

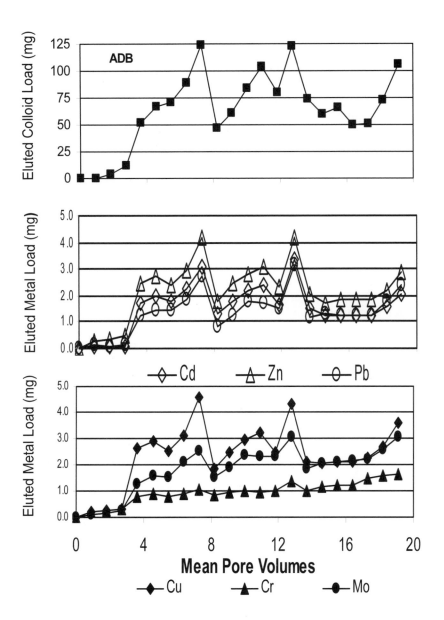

Fig. 7.5. Breakthrough curves for colloid and metal elution in the presence of ADB colloids.

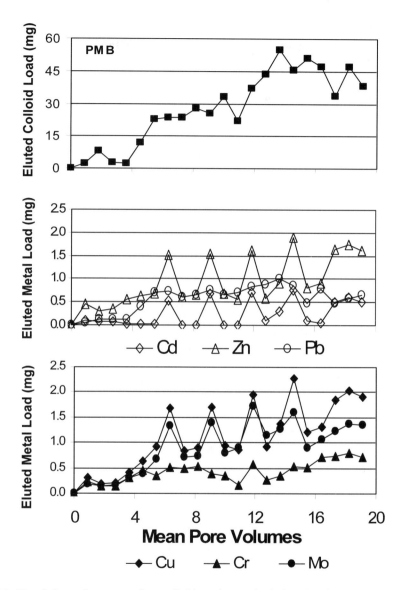

Fig. 7.6. Breakthrough curves for colloid and metal elution in the presence of PMB colloids.

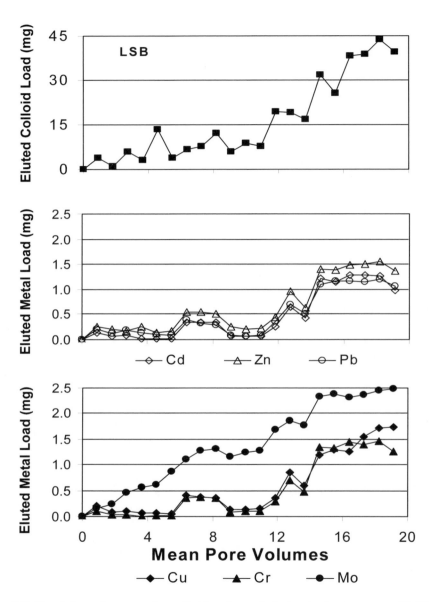

Fig. 7.7. Breakthrough curves for colloid and metal elution in the presence of LSB colloids.

Copper Elution

In spite of the low overall affinity of all colloids for Cu, total Cu elution through all soils combined, along with that of Zn and Mo, was among the highest. This underscores the dynamic nature of transport processes in real soil systems and the misleading interpretations that can be reached in trying to predict metal mobility and attenuation by relying on metal distribution coefficients estimated from batch experiments alone. The ADB colloids were the most efficient overall Cu elutors (50 %), followed by the PMB (24 %), and the LSB (12 %) colloids (Fig. 7.4). Elution patterns were characterized by considerable asymmetry, correlating well with colloid BTC's (0.85^{**}) and corroborating their association in the transport processes (Figs. 7.5-7.7). A couple of Cu breakthrough maxima approaching 1.0 C/C_0 during the leaching cycle are attributed to detachment and flushing of previously strained colloids carrying an extra Cu load in the eluent. This pattern is typical in soils with considerable preferential transport and/or numerous clogging and flushing cycles due to water pressure buildup (Jacobsen et al., 1997). The high organic matter, surface area, and surface charge of the ADB and PMB colloids, in conjunction with their moderately low suspension pH (Table 7.1) promoted colloid and metal mobility, accounting for 56 % of the total variability in total Cu elution as indicated by multiple regression analysis. In addition, the low ionic strength of the above colloid suspensions would tend to promote double layer expansion and limit colloid attachment to the soil matrices. Soluble Cu accounted for 30-60 % of the total Cu eluted through all soils combined (Fig. 7.4), which is attributed to the fairly low affinities of the colloids for Cu (Table 7.1) and their high organic matter status that induced soluble organic-Cu complex formation (Han and Thompson, 1999; Li and Shuman, 1997). The latter may have been accentuated by the higher molecular weight and hydrophilic nature of the ADB organic complexes, which enhanced Cu-binding and colloid mobility (Han and Thompson, 1999). These organo-metal complexes are fairly non-adsorptive within the soil matrix and may explain the sizeable soluble metal fraction mobilized in this study (McBride et al., 1997). As was the case with other metals, the highest Cu elution occurred in association with ADB colloids through the Maury and Bruno soils (48-85 %) and the PMB colloids through the Woolper soil (55 %). Other sizeable elutions of Cu occurred through the Woolper soil in association with the ADB and LSB colloids (~20 %), the Bruno soil in association with LSB colloids (18 %), and the Maury soil in association with PMB colloids (12 %). The decreasing Cu mobility of the latter associations was consistent with lower colloid elution and greater Cu attenuation by the soil matrices. The remaining soil-colloid associations along with the control treatments (absence of colloids) yielded negligible Cu elutions, suggesting complete attenuation by the soil matrices.

Chromium Elution

Total Cr elution was the lowest among metals, averaging 8-21 % through all soils combined, with the ADB colloids being the most effective Cr carriers (Fig. 7.4). Breakthrough curves were generally not as erratic as with other metals, corroborating the lower association with colloid elution (0.43[*]) (Figs. 7.5-7.7). Increased organic matter, low eluent pH, and low residence time stemming from high soil hydraulic conductivity accounted for 67 % of the variability in Cr elution. These elution patterns are consistent with the Cr sorption affinity trends shown by colloids and soils in this study (except for the LSB colloids) and the findings of Balasiou et al. (2001), reporting lower amounts of specifically sorbed Cr to organic matter fractions of soils and colloids than other metals. In spite of their high sorption affinity, LSB colloids eluted low amounts of Cr. It is likely that the high ionic strength of LSB suspensions and potential $Cr(CO_3)$ precipitation on trapped carbonate colloid particles may have limited Cr mobility through the soils (Madrid and Diaz-Barrientos, 1992). The eluted soluble Cr fraction was very low (< 5 %) accounting for 1/5 to 1/3 of the total eluted Cr, mainly in association with ADB and PMB colloids. The small load of soluble Cr is consistent with its low solubility and immobility at pH > 5.5 (Losi et al., 1994). The elution of soluble Cr in association with the LSB colloids and control treatments was negligible, suggesting nearly complete attenuation by the soil matrices. The highest Cr elutions were observed through the Bruno soil in association with the ADB colloids (38 %). The Woolper soil showed the most consistent Cr elution in association with every colloid (16-20 %), while the Bruno soil also eluted similar amounts of Cr in the presence of LSB colloids. The remaining soil-colloid associations and the control treatments yielded negligible Cr recoveries.

Molybdenum Elution

The elution of Mo in association with the biosolid colloids was unlike the elution of any other metals in the study. While significant amounts of other metals were co-transported sorbed onto the biosolid colloids, the majority of total Mo was eluted in the soluble phase due to its dominant anionic form. Yet, significant attenuation of Mo by the soil matrix also occurred. In spite of the zero sorption affinity of soils and colloids for Mo indicated by the Freundlich adsorption isotherm experiments (Table 7.1), total Mo elution correlated reasonably well with colloid elution (r = 0.55[**]). In fact, up to 50 % of the total Mo transported in the presence of ADB colloids was in the sorbed fraction (Fig. 7.4). Apparently, changes in the ionic environment caused by dynamic interactions between the colloids and the soil matrix during the leaching process induced Mo-colloid associations. The highest Mo elution was observed in association with ADB colloids (40 %), followed by the LSB (30 %) and the PMB (20 %) colloids. For the most part, Mo breakthrough in association with biosolid colloids was fairly symmetrical and increased gradually as leaching continued (Figs. 7.5-7.7). These findings are comparable to those of Papelis (2001), who reported little sorption of anionic metal contaminants in transport studies, with symmetrical breakthrough curves and elution mainly in the soluble phase. The Woolper (30-53 %) and Maury (13-29 %) soils most consistently enhanced colloid-mediated Mo elution for all colloids. However, the maximum Mo elution occurred in the Bruno soil in the presence of the ADB colloids (57 %). Multiple regressions suggested a positive influence by OM, carbonate, and hydraulic conductivity increases, which explained 55-66 % of the variability in total and soluble Mo elution. While the carbonate influence can be explained by the accompanied pH increase that enhances Mo mobility, the role of OM must be associated with bridging effects through other cations. In contrast, Fe-oxides and low pH were the dominant factors controlling the elution of sorbed Mo, explaining 61 % of the variability (Brinton and O'Connor, 2003). A detailed identification and characterization of the different metal species is beyond the scope of this chapter. The relatively high efficiency of the LSB colloids to mediate Mo transport more than any other metal is probably due to the presence of carbonates and the high suspension pH, which induces considerable Mo solubilization. In stark contrast, none of the control treatments yielded > 5 % Mo recoveries, suggesting nearly complete attenuation, in spite of the anionic metal form.

7.2.3 Potential Transport Mechanisms

The strong correlation between total metal elution and colloid elution ($R^2 = 0.68^{**}$) and the good agreement between metal and colloid elution patterns implies that a large portion of the metals eluted in the presence of migrating colloids were sorbed or attached onto their surfaces. This was also corroborated from the good correlations between metal transport and colloid charge or specific surface area ($R^2 = 0.62-0.66^{**}$). In spite of the inconsistent affinities of different metals for colloid surfaces vs. the soil matrix and the significant attenuation of transported metals by the soil matrix, an ample metal load was carried by the colloids. This is consistent with the

findings of other investigations emphasizing the potential of mobile colloids as contaminant carriers in subsurface soil environments (Karathanasis, 1999; Denaix et al., 2001). However, the presence of colloids enhanced the elution of both sorbed and soluble metal loads in this study.
Although the so called "soluble metal load" is subject to the scrutiny of the
operational definition of the eluent fraction passing through a < 0.2 μm
membrane filter, XRD and SEM analyses of the eluents did not reveal any
traces of colloids in the filtrate. The correlation coefficient between the
eluted soluble metal fraction and the colloid fraction was particularly high
for Cd, Cu and Zn ($R^2 = 0.76^{**}$ to 0.41^{*}) and explained about 26 % of the
overall total metal elution. In contrast, essentially 0 elution of the metal
soluble fraction occurred in the absence of colloids (control treatments),
suggesting complete attenuation by the soil matrix. This suggests that mobile colloids in the conducted experiments may have played a dual role in
the metal transport process by acting as carriers and facilitators. While the
role of colloids as metal carriers is envisioned through surface chemisorption and perhaps co-precipitation mechanisms, particularly for carbonate-
containing colloids, their role as facilitators requires further documenta-
tion. One possible mechanism contributing to the transport of elevated
soluble metal loads in the presence of colloids, is certainly "complexation"
and "ligand formation" between metals and dissolved organic carbon
(DOC) constituents associated with the biosolid colloids. Metal speciation
simulations suggested that between 20-60 % of the eluted soluble metal
fraction was in the form of organic ligands, while ion specific electrode
analysis of selected metals indicated a range between 35 and 50 %. The
remaining soluble metal fraction eluted in the presence of biosolid colloids
could have been transported via the action of physical or chemical exclusion mechanisms. While these mechanisms can only be speculated in this
study, their functionality has been documented in a variety of physical and
chemical processes, including transport phenomena. Examples of physical
exclusion processes that may result in elution of extra soluble metal loads
include: (a) pore clogging by migrating colloids that block reactive soil
matrix sites within the blocked pore from interacting with soluble or col-
loid-sorbed metals; and (b) increased collisions between soluble metal species and colloid particles during the transport process may decrease attachment or interaction opportunities with soil pore reactive sites.
Examples of chemical exclusion mechanisms contributing to higher soluble metal elution loads include: (a) inactivation of reactive soil pore wall
sites by metal-colloid entities sharing metal ionic bonds, therefore exclud-
ing the sites from interactions with soluble metal species; and (b) soluble
ionic metal species associated with the diffuse double layer of mobile col-

loids may bypass attractive forces of soil matrix reactive sites due to their high coordination and kinetic energy.

7.3 Summary and Conclusions

The potential of biosolid colloids to transport metals associated with organic-waste amendments through subsurface soil environments was investigated with leaching experiments involving undisturbed soil monoliths. The results of the study clearly demonstrated the potential of colloids derived from biosolid wastes to induce the migration of cationic (Cd, Zn, Cu, Pb, Cr), as well as anionic metals (Mo) in subsurface soil environments. Based on mass balance estimates, metal elution in the presence of biosolid colloids was enhanced up to 3 orders of magnitude over that of control treatments without colloids, depending upon colloid and soil type. Metal loads eluted through the three soils in the presence of biosolid colloids increased in the order $Cu = Mo = Zn > Pb = Cd > Cr$, with the ADB colloids transporting nearly twice as much metal load as the PMB colloids and more than 4 times the load of LSB colloids. Metal mobility was strongly correlated with colloid mobility, and was significantly affected by solution pH changes, colloid surface area and charge, soil macroporosity, and OM content. The presence of colloids enhanced the elution of sorbed and soluble metal loads for both cationic and anionic metal forms, suggesting a dual role in the transport process as carriers and facilitators. Although the majority of the eluted metal load was colloid-bound, between 4 % (Pb, Cr) and 60 % (Mo) was soluble. Metal-organic complex formation, either directly (for cationic metals) or indirectly (through bridging for anionic metals) was considered to be the primary cause for soluble metal load increases, but exclusion mechanisms may have also contributed. Metal sorption affinity differences between soils and biosolid colloids were insufficient in predicting metal elution potential, emphasizing the inadequacy of batch experiments alone in assessing the dynamic nature of contaminant transport processes. Metal elution was generally enhanced by decreasing pH and colloid size, and increasing organic carbon content. Soils with high organic matter content and increased macroporosity induced greater metal elution. Breakthrough curves were mostly irregular, showing several maxima and minima as a result of preferential macropore flow and multiple clogging and flushing cycles.

The findings of this study explain the important role of mobile biosolid colloids in explaining metal mass balance losses and the necessity for considering their contribution as transport agents in contaminant transport modelling approaches. This is particularly critical in areas where heavy applications of biosolids as soil amendments may increase the risk for soil and groundwater contamination.

7.4 Outlook

Land application of biosolid wastes as soil conditioners or nutrient sources will continue to rise in the next decade throughout the world. In spite of tighter environmental regulations, contaminants associated with these wastes pose a great risk to groundwater quality through colloid mobilization and transport processes. Existing regulations are still based on soluble metal load mobilization-retention estimates without accounting for biosolid colloids as potential vectors of metal contamination. Therefore, a better understanding of the interactions between active biosolid colloid surfaces and metal species released during rainfall events is warranted. Furthermore, elucidating the mechanistic pathways involved in colloid-metal associations during the transport processes is critical in developing metal migration predictive models as well as implementing effective prevention and remediation strategies. These processes are particularly important for relatively vulnerable ecosystems, such as disturbed lands due to mining activities or karst areas receiving heavy applications of organic wastes, either intentionally or incidentally, that increase considerably the risk for groundwater pollution. In spite of the fact that biosolid colloids constitute some of the most potent vectors of metal and other contaminant transport in these environments, very little laboratory scale research has been done to date to document their potential. While these experiments are conducive to deducing colloid-metal association and potential transport mechanism information, field experiments are essential in order to understand realistic scales of potential contamination under natural spatial and temporal variability conditions. Furthermore, predictive models developed from laboratory experiments need to be validated and the efficiency of prevention and remediation strategies need to be tested under field conditions. Hopefully, this type of information will enable us to develop more realistic environmental protection regulations and safer waste disposal and management practices.

References

Amrhein C, Mosher PA, Strong JE (1993) Colloid-assisted transport of trace metals in roadside soils receiving deicing salts. Soil Sci Soc Am J 57:1212-1217

APHA (1998) Standard methods for the examination of water and wastewater. 20th ed 4500-Cl E. Automated Ferricyanide Method. APHA-AWWA-WPCF

Balasiou CF, Zagury GJ, Deschenes L (2001) Partitioning and speciation of chromium, copper, and arsenic in CCA-contaminated soils: influence of soil composition. Sci Tot Environ 280:239-255

Baveye P, McBride MB, Bouldin D, Hinesly TD, Dahdoh MSA, Abdel-Sabour MF (1999) Mass balance and distribution of sludge-borne trace elements in a silt loam soil following long-term applications of sewage sludge. Sci Total Environ 227:13-28

Bouma J (1991) Influence of soil macroporosity on environmental quality. Adv. Agron 46:1-37

Brinton SR, O'Connor GA (2003) Sorption of Mo in soils field-equilibrated with biosolids. Commun Soil Sci Plan Anal 34:1331-1346

Buchter B, Hinz C, Gfeller M, Fluhler H (1996) Cadmium transport in an unsaturated stony subsoil monolith. Soil Sci Soc Am J 60:716-721

Camobreco VJ, Richards BK, Steenhuis TS, Peverly JH, McBride MB (1996) Movement of heavy metals through undisturbed and homogenized soil columns. Soil Sci 161:740-750

Degueldre C, Grauer R, Laube A (1996) Colloid properties in granitic groundwater systems. II. Stability and transport study. Applied Geochem 2:697-710

del Castillo P, Chardon WJ, Salomons W (1993) Influence of cattle-manure slurry application on the mobility of Cd, Cu, and Zn in a manured, acidic, loamy-sand soil. J Environ Qual 50:689-697

Denaix L, Semlali RM, Douay F (2001) Dissolved and colloidal transport of Cd, Pb, and Zn in a silt loam soil affected by atmospheric industrial deposition.. Environ Pollut 113:29-38

Gove L, Cook CM, Nicholson EA, Beck AJ (2001) Movement of water and heavy metals (Zn, Cu, Pb, and Ni) through sand and sandy loam amended with biosolids under steady-state hydrological conditions. Biores Technol 78:171-179

Han N, Thompson ML (1999) Copper-binding ability of dissolved organic matter derived from anaerobically digested biosolids. J Environ Qual 28:939-944

Hettiarachchi GM, Ryan JA, Chaney RL, La Fleur CM (2003) Sorption and desorption of Cd by different fractions of biosolid-amended soils. J Environ Qual 32:1684-1693

Jacobsen OH, Moldrup P, Larsen C, Konnerup L, Petersen LW (1997) Particle transport in macropores of undisturbed soil columns. J Hydrol 196:185-203

Kaplan DI, Bertsch PM, Adriano DC, Miller WP (1993) Soil-borne mobile colloids as influenced by water flow and organic carbon. Environ Sci Technol 27:1192-1200

Karathanasis AD (1999) Subsurface migration of copper and zinc mediated by soil colloids. Soil Sci Soc Am J 63:830-838

Karathanasis AD, Hajek BF (1982a) Revised methods for rapid quantitative determination of minerals in soil clays. Soil Sci Soc Am J 46:419-425

Karathanasis AD, Hajek BF (1982b) Quantitative evaluation of water adsorption on soil clays. Soil Sci Soc Am J 46:1321-1325

Karathanasis AD, Ming DA (2002) Colloid-mediated transport of metals associated with lime-stabilized biosolids. p 49-63. In: Violante A, Huan PM, Bollag JM, Gianfreda L (ed.) Soil Mineral-Organic Matter-Microorganism Interactions and Ecosystem Health. Developments in Soil Science 28A, Elsevier Science, Amsterdam, The Netherlands

Kretzschmar R, Robarge WP, Amoozegar A (1995) Influence of natural organic matter on transport of soil colloids through saprolite. Water Resour Res 31:435-445

Lamy I, Bourgeois S, Bermont A (1993) Soil Cd mobility as a consequence of sewage-sludge disposal. J Envir Qual 22:731-737

Li Z, Shuman LM (1997) Mobility of Zn, Cd, and Pb in soils as affected by poultry litter extract-I. Leaching in soil columns. Environ Pollut 2:219-226

Linden DR (1995) Agricultural utilization of sewage sludge: A twenty year study at the Rosemount Agricultural Experiment Station (Station Bulletin 606-95). St. Paul, MN: Minnesota Agricultural Experiment Station, University of Minnesota

Logan EM, Pulford ID, Cook GT, Mackenzie AB (1997) Complexation of Cu^{2+} and Pb^{2+} by peat and humic acid. Eur J Soil Sci 48:685-697.

Losi ME, Amrhein C, Frankenberger WT (1994) Factors affecting chemical and biological reduction of hexavalent chromium in soil. Environ Tox Chem 13:1727-1735

Madrid L, Diaz-Barrientos E (1992) Influence of carbonate on the reaction of heavy metals in soils. J Soil Sci 43:709-721

McBride MB (1989) Reactions controlling heavy metals solubility in soils. Adv. Soil Sci 10:1-57

McBride MB, Richards BK, Steenhuis T, Russo JJ, Suave S (1997) Mobility and solubility of toxic metals and nutrients in soil fifteen years after sludge application. Soil Sci 162:487-500

McCarthy JF, Zachara JM (1989) Subsurface transport of contaminants. Environ Sci Technol 23:496-502

McGrath SP, Lane PW (1989) An explanation for the apparent losses of metals in a long-term field experiment with sewage sludge. Environ Pollut 60:235-256

Neal RH, Sposito G (1986) Effects of soluble organic matter and sewage sludge amendments on cadmium sorption by soils at low cadmium concentrations. Soil Sci 142:164-172

NRCS (1996) Soil survey laboratory methods manual. Soil survey investigations report No. 42. Version 3.0, USDA, National Soil Survey Center, Lincoln, NE

Papelis C (2001) Cation and anion sorption on granite from the project Shoal Test Area, near Fallon, Nevada, USA. Adv Environ Res 5:151-166

Persicani D (1995. Analysis of leaching behavior of sludge-amended metals in two field soils. Water Air Soil Pollut 83:1-12

Ritter A, Munoz-Carpena R, Regalado CM, Vanclooster M, Lambot S (2004) Analysis of alternative measurement strategies for the inverse optimization of the hydraulic properties of a volcanic soil. J Hydrol 295:124-139

Ryan JN, Gschwend PM (1994) Effect of solution chemistry on clay colloid release from an iron oxide-coated aquifer sand. Environ Sci Technol 28:1717-1726

Scancar J, Milacic R, Strzar M, Burica O, Bukovec P (2001) Environmentally safe sewage sludge disposal: the impact of liming on the behavior of Cd, Cr, Cu, Fe, Mn, Ni, Pb, and Zn. J Environ Eng 3:226-231

Colloids and Contaminants, Engineering Aspects

8 Ion - Colloid - Colloid Interactions

Willem H. Van Riemsdijk, Liping Weng, Tjisse Hiemstra

Department of Soil Quality, Wageningen University, P.O. Box 8005, 6700 EC, Wageningen, The Netherlands

8.1 Introduction

8.1.1 NOM and Metal (Hydr)Oxides

Natural organic matter (NOM) and metal (hydr)oxides are recognized to be very important geo-colloids due to their omnipresence in nature and due to their high affinity for nutrients and pollutants. These colloids are not only important in soils but are also in sediments, ground water and fresh and marine surface waters (McCarthy and Zachara 1989, Buffle et al. 1998). Humic acids (HA) and fulvic acids (FA) are important reactive components of NOM with respect to cation binding (Thurman 1985, Stevenson 1994). The modern view of the nature of humic substances is that they can form a three-dimensional cross-linked structure and have a size in the nm range (Hayes et al. 1989, Stevenson 1994, Balnois et al. 1999, Duval et al. 2005). Under ambient conditions in the natural environment, humic substances are generally negatively charged. Metal (hydr)oxides are often of colloidal size mineral particles, which are characterized by a surface charge and potential that is strongly dependent on the pH and the chemi-sorption of other components like phosphate. Metal (hydr)oxides bind chemically both cations and anions like Pb^{2+}, Cd^{2+}, PO_4^{3-} SO_4^{2-}, AsO_4^{3-} and humic substances. They also react with neutral species like H_4SiO_4 and H_3AsO_3.

8.1.2 Surface Complexation Models

The interactions between colloids and various ions have been intensively studied. Much knowledge has been gained and advanced surface complexation models (SCM) have been developed to describe the binary system of ion binding to individual type colloids. Ion binding to humic substances is a very complicated process due to the combination of chemical heterogeneity, electrostatic interaction, variable stoichiometry of the binding and the competition between various cations and protons for the reactive binding sites. Extensive research over more than a decade has resulted in the well accepted NICA-Donnan model for ion binding that can handle this complexity (De Wit et al. 1990, De Wit et al. 1993a, 1993b, Koopal et al. 1994, Benedetti et al. 1995, Benedetti et al. 1996, Kinniburgh et al. 1996, Kinniburgh et al. 1999, Milne et al. 2001, Milne et al. 2003, Koopal et al. 2005). In the same time period model V and VI have been developed by Tipping (Tipping and Hurley 1992, Tipping 1994, 1998, 2002) that can also describe competitive ion binding to natural humic materials.

Our knowledge on the interaction between ions and metal (hydr)oxides has also increased enormously. The advances have become possible because a lot of detailed spectroscopic work has been done that gives insight in the structure and nature of the adsorbed species (Parfitt et al. 1976, Rochester and Topham 1979, Hayes et al. 1987, Tejedor & Tejedor and Anderson 1990, Combes et al. 1992, Dent et al. 1992, Manceau et al. 1992, Waychunas et al. 1993, Manceau and Charlet 1994, Spadini et al. 1994, Hug 1997, Rose et al. 1997, Bargar et al. 1999, Collins et al. 1999, Connor et al. 1999, Ostergren et al. 1999, Parkman et al. 1999, Peak et al. 1999, Randall et al. 1999, Wijnja and Schulthess 1999, Bargar et al. 2000, Sahai et al. 2000, Arai and Sparks 2001, Redden et al. 2001, Villalobos and Leckie 2001, Sherman and Randall 2003, Spadini et al. 2003, Kim et al. 2004). From the modelling point of view the challenge is to develop sound thermodynamic models that make use of the knowledge gained from spectroscopic studies. The CD-MUSIC model has been developed for this purpose (Hiemstra et al. 1989a, Hiemstra et al. 1989b, Hiemstra and Van Riemsdijk 1996, Hiemstra et al. 1996, Hiemstra and Van Riemsdijk 2006b). The model is based on the structure of the mineral surface, the structure of the adsorbed species and the structure of the electrostatic potential profile in the vicinity of the mineral-water interface.

8.1.3 Ion Speciation in Soils, Sediments and Water

Field samples always contain a mixture of several adsorptive surfaces. For instance, colloidal particles of NOM, clay alumino-silicates and metal (hydr)oxides are all considered important reactive surfaces that bind heavy metals. Despite advances in the development of sophisticated models for individual type of colloidal particles, application of these models to natural environmental samples like soils and sediments is hampered by the complexity of the real systems. One of the difficulties is to estimate the amount of reactive surface of each type of sorbent. Another complication arises due to the interactions between the colloidal particles. Attempts have been made to apply the surface complexation models for individual adsorptive surfaces to understand and predict ion speciation in natural samples (Weng et al. 2001a, Weng et al. 2002, Cances et al. 2003, Lumsdon 2004, Dijkstra et al. 2004a, Schroder et al. 2005). These studies focused mainly on the behaviour of cations. For simplicity, it was assumed in these studies that there is no effect of the particle interaction on the ion binding properties of the particles. Despite of these simplifications, rather promising results have been obtained so far for the metal ions Cu and Cd. For Ni, Zn and Pb, the model predictions are sometimes adequate, sometimes quite poor. For oxyanions like phosphate and arsenate metal (hydr)oxides and edge faces of clay minerals are the most important reactive surfaces in nature and competition with natural organic matter probably has a significant effect on the binding of these anions. At present we can not handle this problem in a satisfactory way.

8.1.4 Interactions between Colloidal Particles

Besides the interactions with cations and anions, colloidal particles interact with each other as well. Humic substances adsorb readily onto oxide minerals (Tipping 1981, Davis 1982, Murphy et al. 1992, Gu et al. 1994, Spark et al. 1997, Vermeer and Koopal 1998, Filius et al. 2000, Saito et al. 2004). The adsorption of humics will influence both the solubility and mobility of NOM and the surface properties of oxides. The interactions between particles will also affect the adsorption of ions to both particles (Robertson and Leckie 1994, Zachara et al. 1994, Murphy and Zachara 1995, Vermeer et al. 1999, Christl and Kretzschmar 2001a, Weng et al. 2005). The final outcome with respect to both the humics adsorption to the oxides and the ion binding to the adsorption complex is the net result of various interactions, and depends on the nature of the surface and the humics, the type of ions and factors such as pH and ionic strength (Weng et al. 2005). To be able to predict the adsorption of both small ions and organic particles to the oxide surface in a thermodynamic consistent manner is one of the challenges that have to be solved before the ion-binding models for oxides can be applied to environmental systems in a useful manner.

8.1.5 Modelling Particle Adsorption

Modelling the adsorption of variable charge nano-particles like fulvic and humic acids is an important fundamental area of research for colloid and physical chemistry. A simplified approach is to treat the nano-particles as small ions. Such an approach has been applied to the adsorption of humics (Ali and Dzombak 1996, Filius et al. 1997a, Karltun 1998, Filius et al. 2000). This type of modelling approach has been referred to by Filius *et al.* (Filius et al. 2003) as "discrete modelling". In this approach, a step-wise protonation reaction is formulated to describe the protonation reactions of humics in the solution phase. For the adsorbed humic molecules, *a priori* assumptions are made with respect to a limited number of hypothetical humic surface species. For macromolecules such as humics, the number of reactive ligands present on each particle is large. Therefore the number of possible species that the molecules can form in both phases is huge. By reducing the number of species, as has been done in the "discrete modelling" for humics, the speciation of the adsorbed particles calculated is chemically and physically most probably not very realistic.

Recently the concept of the LCD (Ligand and Charge Distribution) model has been proposed (Filius et al. 2001, Filius et al. 2003, Weng et al. 2005), which combines the CD-MUSIC model with the NICA model that allows for the description of the adsorption of humic substances on metal (hydr)oxides. In comparison to other models developed for polyelectrolyte adsorption, the advantage of the LCD model is that it allows for the incorporation of details that are already present in both the CD-MUSIC and the NICA model, *i.e.* the chemical heterogeneity, variable reaction stoichiometry, competition of binding, and the structure of the electrostatic potential profile in the vicinity of the mineral-water interface. The LCD model framework is a promising modelling tool that integrates the ion binding models for both humic substances and metal (hydr)oxides. This approach is an important step in the application of geochemical models to environmental systems.

In this chapter, an overview will be given with respect to recent modelling developments for ion-colloid-colloid interactions and the attempts that have been made to apply these (semi)mechanistic models to natural samples. First, we will briefly explain two representative advanced models developed for ion binding to respectively NOM and metal (hydr)oxides, *i.e.* the NICA-Donnan model and the CD-MUSIC model. It has been shown that these models can describe ion binding to purified humics or synthetic metal (hydr)oxides well. The models have to some extent also been tested with respect to their predictive capability of competitive ion binding. An important practical reason of using a binding model that is as close to physical reality as possible is that it is more likely to give reliable predictions for competitive binding. In natural systems competition of variable nature and intensity will always take place. Secondly, we will discuss some of the attempts that have been made to apply these models in geochemical equilibrium modelling to understand or predict ion speciation in soil samples. In these models, it has been assumed that the interactions between particles do not affect ion binding to the mixture (linear additivity). Thereafter, we will address the phenomena related to humics adsorption to metal (hydr)oxides and the effects on ion binding. The recent modelling development for humics adsorption to metal (hydr)oxides, the LCD model, will be presented, and in the end of the chapter the need for future research will be discussed.

8.2 Ion Binding to Natural Organic Matter

8.2.1 Ion Binding to NOM

Natural organic matter (NOM) is omnipresent in soils, sediments, surface and groundwater and it consists of a complex mixture of different types of molecules. Depending on its solubility in acid and base solutions, NOM can be fractionated into fulvic acids (FA), humic acids (HA) and humin (Hayes et al. 1989, Stevenson 1994). Most of the fulvic and humic acids fall within the colloidal range, and have a higher density of reactive functional groups than the humin fraction. The humic and fulvic acid fractions are the most reactive fractions of NOM in terms of metal ion binding. The fulvic and humic acids are operationally defined fractions and each category will be a mixture of ill-defined molecules with similar properties. The heterogeneous nature of the fulvic and humic materials, the presence of electrostatic interactions, the different binding stoichiometry for different ions and the challenge to predict competitive metal ion binding makes it quite difficult to develop models that can handle this complexity successfully (Koopal et al. 2005). The earlier work on modelling ion binding to NOM was reviewed by Buffle (Buffle 1988) and an overview of more recent developments can be found in the review of Dudal & Gerard (Dudal and Gerard 2004) and the book of Tipping (Tipping 2002).

8.2.2 WHAM and NICA-Donnan Model

Intensive research in the last decade has led to the development of advanced models for ion binding to humic materials. The most accepted models in this category are the WHAM (Windermere humic aqueous acid, model IV-VI) (Tipping and Hurley 1992, Tipping 1994, 1998, 2002) and the NICA-Donnan model (De Wit et al. 1990, De Wit et al. 1993a, 1993b, Koopal et al. 1994, Benedetti et al. 1995, Benedetti et al. 1996, Kinniburgh et al. 1996, Kinniburgh et al. 1999, Milne et al. 2001, Milne et al. 2003, Koopal et al. 2005). Both models take the heterogeneity, competition, variable reaction stoichiometry and electrostatic interactions into account (Koopal et al. 2005). Humic substances are "soft" particles that are to some extent penetrable for the co- and counter-ions that neutralize the charge of the humic colloids. Because of the ill defined nature of the humic colloids often a simplified approach is used to account for the electrostatic effects on ion binding to humics. The Donnan model assumes a constant homogenous electrostatic potential in the humic phase for given solution conditions like pH, salt level *etc*. Both WHAM and NICA-Donnan model have adopted a Donnan approach to separate the "intrinsic" affinity from the electrostatic effects. However, the implementation of the Donnan approach is somewhat different for the two model approaches. The total amount of ions bound to the humic molecules includes both the ions that bind specifically to the reactive groups of the organic matter as well as the amount bound electrostatically in the Donnan part of the model.

The WHAM and NICA-Donnan model have followed different approaches to deal with chemical heterogeneity and variable stoichiometry. In the WHAM model, the distribution of the intrinsic affinity is treated as a discrete distribution. Four types of carboxylic sites and four types of phenolic sites are assumed with the total density of the carboxylic sites taken as two times that of the phenolic sites. Each of the four types of the carboxylic or phenolic sites has an equal density and its affinity is distributed around a mean affinity value for each site type. The WHAM model deals with the variable stoichiometry by explicitly allowing for the formation of bidentate and tridentate complexes for certain metal ions and reactive sites. This approach also involves a series of more or less arbitrary assumptions in order to make it a practical approach. The NICA model uses another philosophy to deal with the heterogeneity of the affinity of the binding sites and the reaction stoichiometry. Differing from the WHAM model, the NICA model assumes that the affinity of the carboxylic and phenolic type of sites has a continuous distribution following one of the Sips' distribution functions (Sips 1948). This assumption has the big advantage that an analytical competitive binding equation results. The competition between ions and the variable stoichiometry of the binding are accounted for by using a competitive Hill type equation as the local isotherm. Combination of this competitive isotherm with the Sips distribution function results in the NICA model, which is a relatively simple very elegant equation:

$$N_{\mathrm{i}} = N_{\max} \frac{n_i}{n_{\mathrm{H}}} \theta_i = Q_{\max} \frac{n_i}{n_{\mathrm{H}}} \frac{(\tilde{K}_i C_i)^{n_i}}{\sum (\tilde{K}_i C_i)^{n_i}} \frac{\left\{\sum (\tilde{K}_i C_i)^{n_i}\right\}^p}{1 + \left\{\sum (\tilde{K}_i C_i)^{n_i}\right\}^p}, \qquad (8.1)$$

where $\tilde{K}_i$ is the mean affinity constant, C_i is the local concentration of component i. The parameter n_i ($0 < n_i \leq 1$) is related to the degree of correlation between the affinity distribution of the proton and the metal ion i (Rusch et al. 1997, Kinniburgh et al. 1999) and is also related to the stoichiometry of the binding reaction of the ion. The width of the affinity distribution, indexed as p ($0 < p \leq 1$), represents the intrinsic heterogeneity of the ligands. Different ions have different values for the parameter n_i. Protons are used as the probe ions and the total site density ($N_{\max}$) is defined as the maximum amount of protons that can be bound. The stoichiometry of other ions relative to that of proton can be derived from the ratio of n_i to n_{H}. Note that the NICA model simplifies to a competitive Langmuir model when all n_i values and p are equal to one.

8.2.3 Application of the NICA-Donnan Model to Purified FA and HA

The NICA-Donnan model has been applied to describe proton and metal ion binding to purified fulvic and humic acids (Benedetti et al. 1995, Kinniburgh et al. 1999, Pinheiro et al. 1999, 2000, Christl and Kretzschmar 2001b, Christl et al. 2001c). An overview of these studies can be found in the paper of Koopal et al. (Koopal et al. 2005). The results obtained are quite good (Tipping 2002, Dudal and Gerard 2004). Using data collected from the literature, Milne et al. (Milne et al. 2001, Milne et al. 2003) have derived generic NICA-Donnan model parameter sets for fulvic acids and humic acids. The generic constants describe the behaviour of an 'average' fulvic or humic acid based on literature data. In Fig. 8.1, the model predictions of the charging behaviour of Strichen FA and Tongversven forest HA and Cu binding to Strichen FA are compared to experimental data. Fig. 8.1 shows that for Strichen FA, the charging/pH curve is described well with the NICA-Donnan model using the generic model parameters. However, for Tongversven forest HA, the generic model parameters lead to a prediction that overestimates the change of charge with pH. The data can be very well described by the model when the model parameters are optimized for this particular dataset (Milne et al. 2001). This behaviour is not unexpected because of the variation in the nature of humic acids from different origin. Fig. 8.1 also shows that for Cu binding to

Strichen FA, the generic parameters underestimate the binding systematically. A very good model description for the Cu data can again be obtained by using model parameters that are optimized for this dataset. The parameters used to calculate the curves in Fig. 8.1 are listed in Tables 8.1 and 8.2.

Table 8.1. NICA-Donnan model parameters for fulvic acid

	Generic (Milne et al. 2001, Milne et al. 2003)		Specific (Milne et al. 2001)	
	Carboxylic site	Phenolic site	Carboxylic site	Phenolic site
Q_{max}(mol/kg)	5.88	1.86	5.67	1.53
m	0.38	0.53	0.45*	0.53*
p	0.59	0.70	0.68	0.70
$\log \widetilde{K}_H$	2.34	8.60	2.77	8.60
n_H	0.66	0.76	0.66	0.76
$\log \widetilde{K}_{Cu}$	0.26	8.24	1.23*	8.01*
n_{Cu}	0.53	0.36	0.54*	0.46*
b	0.57		0.66	

*This work

Table 8.2. NICA-Donnan parameters for humic acid

	Generic (Milne et al. 2001)		Specific (Milne et al. 2001)	
	Carboxylic site	Phenolic site	Carboxylic site	Phenolic site
Q_{max}(mol/kg)	2.30	3.92	2.40	2.39
m	0.50	0.26	0.49	0.26
$\log \widetilde{K}_H$	2.21	8.98	2.49	8.00
b	0.30		0.31	

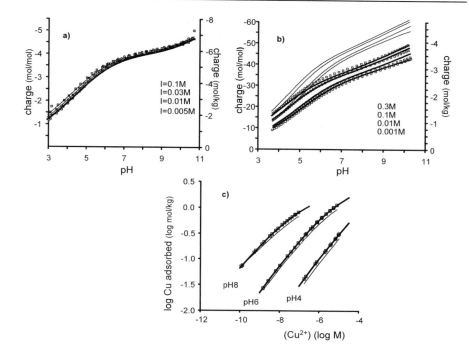

Fig. 8.1 Charging/pH curve of Strichen FA (a), Tongversven forest HA (b), and Cu binding to Strichen FA (c). Thick lines are model prediction using specific parameters, thin lines are model prediction using generic parameters. Symbols are experimental data. For the charge on FA and HA, the value is given in both mol/mol and mol/kg. In the unit transformation, the molar mass used for FA and HA is respectively 683 g/mol and 13,189 g/mol.

8.3 Ion Binding to Metal (Hydr)Oxides

8.3.1 Modelling Ion Binding to Metal (Hydr)Oxides

In terms of their abundance, the (hydr)oxides of iron, aluminium and manganese are the most important metal (hydr)oxide minerals in soils, sediments and water. Depending on environmental conditions such as pH, metal (hydr)oxides can be positively charged, negatively charged or uncharged. They are important colloidal particles in the environment and can bind various cations and anions. Experimental data of charging and ion adsorption to metal (hydr)oxides can be described in various empirical ways, using a partitioning coefficient, simple isotherm equations and conditional binding constants (Dzombak and Morel 1990). Metal (hydr)oxides have an electrostatic surface potential that varies strongly with pH. The electrostatic effect on the binding of ions to oxides is much stronger than it is for NOM. On the other hand the effect of the chemical heterogeneity of the binding sites on the ion binding is much less than it is for NOM.

A series of surface complexation models (SCM) have been developed and applied, such as the Constant Capacitance model (Atkinson et al. 1967, Schindler and Kamber 1968, Sigg and Stumm 1981, Goldberg and Sposito 1984, Lovgren et al. 1990), the Basic Stern model (Yates et al. 1974, Westall and Hohl 1980, Hiemstra et al. 1987, Hiemstra and Van Riemsdijk 1991, 1996, Venema et al. 1996a, Borkovec 1997, Schudel et al. 1997, Felmy and Rustad 1998, Lutzenkirchen 1998, Christl and Kretzschmar 1999) and the Triple Layer model (Yates et al. 1974, Yates 1975, Davis et al. 1978, Hayes and Leckie 1986, Sverjensky 1994, Sahai and Sverjensky 1997, Rudzinski et al. 1999, Villalobos and Leckie 2001, Sverjensky 2005), the Generalized Two Layer model (Dzombak and Morel 1990) and the Variable Charge -Variable Potential model (Bowden et al. 1977, Barrow and Bowden 1987). Most of the models assume a homogeneous surface with ill-defined surface sites and most models use a point charge approach for the adsorbed ions. Since one will adjust model parameters to describe a set of adsorption data for a given model choice, it means that the electrostatic contribution to the binding may differ considerably between different model options. This has practical significance for application to more complex environmental systems because the interaction between adsorbing ions on mineral surfaces like metal-oxides is to a large extent due to electrostatic effects.

These SCM are usually able to describe the experimentally determined adsorption phenomena for simple systems, as long as the number of chosen type of surface species is not constrained. The surface species needed to describe the data will not have *a priori* a direct link with physical realistic species as observed with *in situ* spectroscopy. Important to realize is that the models may lead to very different predictions in case of extrapolation to systems, which were not part of the calibration, as may be the case when competition between various species is relevant. A way to improve SCM is the attempt to incorporate knowledge on the structure of the surface, the structure of surface complexes and the structure of the double layer as available from crystallography, spectroscopy, and other techniques. Ideally, one would like to use a model approach that is not only thermodynamically correct, but that is also as close as possible in agreement with a structural analysis of the surface since the reactive surface sites are the basis for the formulation of all chemical binding to the surface. This approach is not only from a fundamental scientific perspective of importance, it has also practical relevance since the predictive capability of a model approach is expected to be better the closer the model is to physical reality.

8.3.2 MUSIC and CD Model

The MUlti SIte Complexation (MUSIC) model (Hiemstra et al. 1989b) defines surface sites as derived from the structure of the mineral for a given crystal face. The formal charges of the various types of reactive groups are based on the application of the Pauling bond valence concept. This assures a proper bookkeeping of the surface charge as a function of the protonation. At the surface of iron and aluminium (hydr)oxides, one can have surface oxygens with one, two or three bonds to an underlying iron in the mineral. Using a simple Pauling bond valence analysis to assign the formal charge to the various types of surface oxygens for goethite (Hiemstra et al. 1989a, Hiemstra et al. 1996), shows that the formal charge of the non protonated triply, doubly and singly coordinated surface oxygens are respectively -0.5, -1, and -1.5 valence units (v.u.). The application of the MUSIC model to various minerals shows that sites that have a very high or very low value of the formal charge will not be present at the surface to any significant extent. In other words, such surface species are thermodynamically not very stable. For the singly coordinated surface oxygens on iron (hydr)oxides like goethite, the concentration of the $\equiv FeO^{1.5-}$ site will be insignificant and the major surface species are thus the $\equiv FeOH^{0.5-}$ and the $\equiv FeOH_2^{0.5+}$ species, whereas for the triply coordinated oxygen the $\equiv Fe_3O^{0.5-}$ and the $\equiv Fe_3OH^{0.5+}$ are relevant species. The doubly coordinated oxygens have a high affinity for protons, which leads to the formation of a stable uncharged species $\equiv Fe_2OH^0$. In the MUSIC model, the basic charging of the mineral goethite is therefore assumed to be caused by the protonation and deprotonation of the singly ($\equiv FeOH^{0.5-}$) and triply ($\equiv Fe_3O^{0.5-}$) coordinated surface oxygens:

$$\equiv FeOH^{0.5-} + H^+ \leftrightarrow \equiv FeOH_2^{0.5+} \quad \log K_{H,1} \qquad (8.2)$$

$$\equiv Fe_3O^{0.5-} + H^+ \leftrightarrow \equiv Fe_3OH^{0.5+} \quad \log K_{H,2} \qquad (8.3)$$

Note that we have a model approach with one log$K$ per site type and in total two log$K$ values. In the conventional approach one assumes a two-step protonation reaction for one site type (which is the same for all oxides). According to the MUSIC model the conventional approach is in general not very realistic from a physical point of view. Based on the structure analysis, the overall effective site density of goethite is set at 3.45 site/nm^2 for the singly coordinated oxygen and 2.7 site/nm^2 for the triply coordinated oxygen. The (refined) MUSIC model also allows for an *a priori* estimate of the log$K$ values of the various protonation reactions for various minerals. In practice the model for goethite can be simplified *e.g.* by assuming that the proton affinity constants for the two types of reactive sites are equal. The intrinsic proton affinities (log$K_{H,1}$ and log$K_{H,2}$) follow in that case directly from the PZC (point of zero charge) of goethite. A similar approach can be followed for TiO$_2$ (Machesky et al. 2001, Bourikas et al. 2001a, Ridley et al. 2002, Ridley et al. 2004).

The electrostatic models that have been used in combination with the MUSIC model include the Basic Stern and the Three Plane (TP) model. The TP term is used to stress the difference with the TL model approach. Ions that form inner-sphere complexes with the surface have part of their ligands in the surface plane and the part that is not directly involved in the chemical binding is further away from the surface. The fraction of the charge that is directly involved in the chemical binding depends on the co-ordination number (CN) of the ion (*e.g.* 3 for SeO$_3^{2-}$, 4 for PO$_4^{3-}$, and 6 for Cd^{2+}) and on the number of bonds with the surface, i.e. one in case of a mono-dentate surface complex and two in case of a bi-dentate surface complex *etc*. In the CD (Charge Distribution) model, the charge of an adsorbing ion is distributed over two electrostatic planes. It has been shown that the value of the CD parameter for inner-sphere complexes is closely related to the fraction of the ligands that forms bonds with the surface (Rietra et al. 1999, Hiemstra and Van Riemsdijk 2002). For a bi-dentate phosphate surface complex roughly half of the charge of the phosphate ion is attributed to the surface and half to the electrostatic plane that is closest to the surface. In contrast, the bidentate complex of SeO$_3^{-2}$ attributes about 2/3 of the charge to the surface. The advantage of this approach is that the electrostatic energy component to the binding is now a combination of two terms, one for each plane. The CD approach is a practical way to account for the fact that in reality there is a gradient of the electrostatic potential over the adsorbed ion. It is also an advantage that the value of the CD parameter is linked to the structure of the surface complex. The simplest electrostatic model that takes the structure of the inner-sphere surfaces into account is a CD approach with two electrostatic planes, a CD-Basic Stern model. The outer-sphere complexes are in that case still treated as point charges and put at the outermost plane that coincides with the head end of the ddl.

It is interesting to note that recently some authors have started to use a CD approach in combination with a 2 pK-TL model, where the CD value is treated as a fitting parameter (Villalobos and Leckie 2001). They studied bicarbonate adsorption on goethite in terms of the pH dependence of the binding and using IR spectroscopy. Their interpretation of the spectroscopic data was that the surface complex is predominantly a mono-dentate complex, whereas the fitted CD value points more in the direction of a bi-dentate surface complex. Hiemstra et al. (Hiemstra et al. 2004a) reinterpreted the spectroscopic data and showed that the surface complex is most likely a bi-dentate complex, which would mean that the interpretation from both spectroscopic data and the adsorption model are now no longer in conflict with each other. Recently, Bargar et al. (Bargar et al. 2005) have shown in a spectroscopic study that indeed the carbonate species binds as a bidentate inner-sphere complex.

Recently, the charge distribution approach has been applied to electrolyte ions that may adsorb as outer-sphere complexes (Hiemstra and Van Riemsdijk 2004, Hiemstra et al. 2004b, Rahnemaie et al. 2006a, 2006b). In this study, an electrostatic three plane model approach was used. The charge of the electrolyte ions, adsorbed as outer-sphere complexes, was allowed to distribute over the two outermost planes. This enables a test of the position of electrolyte ion charge in the double layer profile. For the test, carefully designed experiments are required in which the surface charge data, obtained in different ionic media, are made internally consistent (Rahnemaie et al. 2006a, 2006b). Very recently the extensive data have been (re)interpreted (Hiemstra and Van Riemsdijk 2006b), treating the electrolyte ions in a classical manner as a single ion charge. The analysis shows that the minimum distance of approach of electrolyte ions differs from the position of the head end of the diffuse double layer. According to Hiemstra and Van Riemsdijk (Hiemstra and Van Riemsdijk 2006b), this separation is due to the alignment of interfacial water molecules, only allowing electrolytes ions to penetrate stepwise into the last few water layers near the surface. The observed double layer picture is in agreement with data from force measurements (Pashley and Israelachvili 1984), and X-ray reflectivity (Catalano et al. 2006). The new model with the additional Stern layer can be called an Extended Stern (ES) layer model (Westall and Hohl 1980).

8.3.3 Applications of MUSIC Model to Proton Adsorption to Metal (Hydr)Oxides

The MUSIC model has been applied to gibbsite (Hiemstra et al. 1987, Hiemstra et al. 1999, Bickmore et al. 2004), boehmite (Jolivet et al. 2004), goethite (Hiemstra et al. 1996), hematite (Schudel et al. 1997, Venema et al. 1998, Hiemstra and Van Riemsdijk 1999b), lepidocrocite (Vermeer and

Koopal 1998), magnetite (Wesolowski et al. 2000, Machesky et al. 2001, Jolivet et al. 2004), silica (Hiemstra et al. 1989b, Hiemstra and Van Riemsdijk 2002), cerium oxide (Nabavi et al. 1993), rutile (Machesky et al. 2001, Bourikas et al. 2001a, Ridley et al. 2002, Ridley et al. 2004) and anatase (Bourikas et al. 2001a). Recently, the model has also been applied to interpret the charging of edges of alumino-silicate clay minerals (Tournassat et al. 2004) and imogolite (Nabavi et al. 1993). The refined MUSIC model has been extended to include the effect of temperature on the proton affinity constants and thus the temperature dependence of the PZC by Machesky et al. (Machesky et al. 2001). The MUSIC model has also been applied successfully to rutile, using calculated *ab initio* bond lengths from quantum chemical calculations and hydrogen bonding information derived from molecular dynamic (MD) simulations (Zhang et al. 2004).

8.3.4. Applications of CD Model to Ion Adsorption to Metal (Hydr)Oxides

The CD model has been applied for metal oxides to the adsorption of inorganic cations like Cd^{2+} (Venema et al. 1996b, Weerasooriya et al. 2002), Ca^{2+} (Rietra et al. 2001b, Weng et al. 2005, Rahnemaie et al. 2006b), Mg^{2+}, Sr^{2+} (Rahnemaie et al. 2006b), Cu^{2+} (Tadanier and Eick 2002) and anions like F^- (Hiemstra and Van Riemsdijk 2000), PO_4^{3-} (Hiemstra and Van Riemsdijk 1996, 1999a, Tadanier and Eick 2002, Antelo et al. 2005), AsO_4^{3-} (Hiemstra and Van Riemsdijk 1999a, Weerasooriya et al. 2004, Antelo et al. 2005, Stachowicz et al. 2006), SeO_3^{2-} (Hiemstra and Van Riemsdijk 1999a), SeO_4^{2-} (Rietra et al. 2001a), SO_4^{2-} (Rietra et al. 2001a, 2001b), MoO_4^{2-} (Bourikas et al. 2001b, Gustafsson 2003), and neutral species like $As(OH)_3$ (Weerasooriya et al. 2003, Stachowicz et al. 2006) and H_4SiO_4 (Gustafsson 2001). In addition, applications are found for the adsorption of small organic acids (Filius et al. 1997b, Geelhoed et al. 1998, Boily et al. 2000), fulvic acid (Filius et al. 2000), organic (m)ethyl As-complexes (Jing et al. 2005). The CD model has also been used in predicting ion binding in soils (Weng et al. 2001a, Schroder et al. 2005).

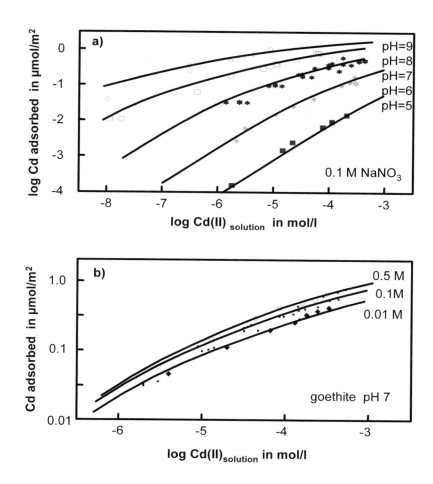

Fig. 8.2. Adsorption isotherms of Cd^{2+} on goethite as function of pH (a) and of ionic strength (b). The adsorption is described assuming bidentate innersphere complexation on the 110 face as observed by EXAFS (Spadini et al. 1994). At the top end of the crystals (021 and 001 faces), high affinity sites are assumed to be present. The lines are calculated with the CD model. The parameters are given by Venema et al. (Venema et al. 1996b)

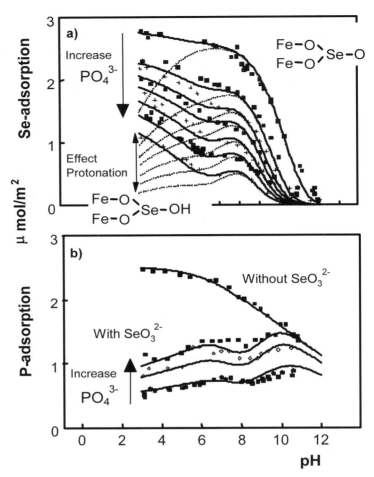

Fig. 8.3. The adsorption of SeO_3^{2-} as a function of pH for goethite systems with a given amount of SeO_3^{2-} but an increasing amount of PO_4^{3-} present (Data of Hingston et al. (Hingston et al. 1971). In Fig. 8.3(b), the corresponding adsorption of PO_4^{3-} is given for the same systems. For comparison, the PO_4^{3-} adsorption without competition of SeO_3^{2-} is also given in Fig. 8.3(b). The ion competition is described with the CD model (Hiemstra and Van Riemsdijk 1999a). For PO_4, the formation of bidentate and protonated bidentate complexes are assumed, as observed with CIR-FTIR (Tejedor & Tejedor and Anderson 1990). The adsorption of SeO_3^{2-} is described assuming bidentate and protonated bidentate complexation. The protonation of adsorbed SeO_3^{2-} is enhanced as result of the presence of negative charge due to the adsorption of PO_4.

An example of the modelling of Cd^{2+} adsorption on goethite as function of concentration, pH, and ionic strength is given in Fig. 8.2. The binding can be described using bidentate innersphere surface complexes as observed by spectroscopy. The major species at the main crystal face has a fitted surface charge attribution (0.6 v.u.) close to the value expected based on complex structure. In the surface complex, two of the six ligands coordinating with Cd^{2+} are common with the surface corresponding to 2/6=1/3 of the charge of Cd^{2+}, i.e. 0.67 v.u. In Fig. 8.3, the adsorption of Se is given in systems with increasing concentrations of PO_4^{3-}. The competitive interaction of both ions can be described well, provided that protonation of the bidentate surface complex is assumed. The charge distribution in the interface was calculated assuming a Pauling distribution of ion charge, i.e. two third of the ligands are common with the surface or -4/3 v.u. is to be attributed.

8.4 Modelling Ion Speciation in Natural Samples

8.4.1 Ion Speciation in Soils, Sediments and Water

To maintain the sustainability of ecological systems such as soil, sediment, groundwater and surface water, it is necessary to make accurate assessment of the toxicity, bioavailability and mobility of various compounds. The behaviour of a compound is dependent not only on its total content, but also on its distribution among various chemical forms (speciation). It is known that the processes of precipitation, adsorption, and complexation are important in controlling ion speciation in the environment. Ion adsorption to various particle surfaces present in the system plays a key role in determining the speciation of the ions in many situations, and thus their bioavailability and mobility.

8.4.2 Empirical Modelling Approaches

Efforts have been made to develop models to describe ion speciation in the environment. Mostly, these models are of empirical nature and consider the system under study as a black box. The solid-solution partitioning coefficient (K_d) for ions can vary many orders of magnitude (Anderson and Christensen 1988, Sauve et al. 2000, Sauve et al. 2003) and people have realized that the K_d value is not a good parameter to use in calculating ion speciation. Other empirical models fit the data to functions such as the Freundlich or Langmuir type of equations (Oakley et al. 1981, Anderson and Christensen 1988, Temminghoff et al. 1994, Temminghoff et al. 1995, Janssen et al. 1997, Elzinga et al. 1999, Sauve et al. 2000, Strawn and Sparks 2000, Degryse et al. 2003, Impellitteri et al. 2003). The goodness of fit of these empirical functions can be improved to a certain extent by relating the amount of adsorption to the amount of the adsorbents, *e.g.* NOM, rather than to the whole soil or sediment (Temminghoff et al. 1997). These functions can describe the calibrated dataset well, but can fail when extrapolation is made to other conditions. The changes of the composition and solution chemistry of the system hamper the application of the empirical relations to a wider range of samples.

8.4.3 Geochemical Equilibrium Calculations

In the last decades, the development of computer codes in the geochemistry field has greatly increased our ability to carry out sophisticated speciation calculations. These computer programs are aimed at solving thermodynamic equations relevant to geosciences and can calculate aqueous complexation, adsorption, precipitation, kinetics and transport. Examples of these codes are: MINTEQA2, PHREEQC, ECOSAT, WHAM and ORCHESTRA (Allison et al. 1991, Keizer and Van Riemsdijk 1994, Tipping 1994, Parkhurst and Appelo 2002, Meeussen 2003). Most of these programs are accompanied with databases that contain common thermodynamic constants such as complexation constants for inorganic aqueous species, solubility products for minerals, *etc*. In the past, due to the lack of appropriate ion adsorption models, the adsorption of ions is calculated in some of these geochemistry codes using empirical functions, which is one of the reasons that have led to unsatisfactory results in applying the geochemical models to real samples (Mouvet and Bourg 1983, Hesterberg et al. 1993). The empirical nature of the adsorption models used also limits the predictive ability of these calculations.

8.4.4 From the Model System to the Field

As has been discussed in Sects. 8.2 and 8.3, more advanced models have become available in the last decade with respect to ion adsorption to geocolloids, such as the NICA-Donnan model for ion binding to NOM and the CD-MUSIC model for ion binding to metal hydr(oxides) (see Sects. 8.2 and 8.3). These SCMs are aimed to represent as close as possible the structure of the particle surface and the physical chemical reality of the adsorbed ion species. It has been shown that these models can describe ion binding to purified or synthetic particles well (see Sects. 8.2 and 8.3). For these models, intensive research has been carried out and model parameters have become available for many ions by model application to experimental dataset of purified HA and FA or synthetic metal (hydr)oxides. The developments of these models together with the availability of model parameters have made it possible to include the (semi)mechanistic ion adsorption models in the geochemical equilibrium calculations.

However, several questions arise when going from the model systems to natural samples. One of the questions is how to determine the amount of reactive sites or reactive surface area of the adsorbents in the natural samples. For instance, soils, sediments and water samples contain always natural organic matter, which is a mixture of many types of molecules. The site density and reactivity of the NOM may differ from one source to another, and may be different from the model material of purified FA and HA. Metal (hydr)oxides such as iron (hydr)oxides may be present in various crystallized forms. The surface area per unit mass of metal (hydr)oxides depends on the size distribution of the mineral particles.

Another complication in making the step from the model to the field is due to the fact that soils, sediments and water are multi-component systems, which means there are more than one type of ions that will compete for the binding sites of the adsorbents. Sometimes, the adsorption of the ions under concern is much stronger than the other ions and the effect of the competition is relatively weak and can be neglected. However, in other cases, the competition effect on the binding of ions of interest cannot be neglected. In these cases, competitive and/or synergistic effects have to be taken into account in order to calculate the speciation with acceptable accuracy. Although the (semi)mechanistic ion binding models like the NICA-Donnan and the CD-MUSIC can handle competitive binding, not all information needed is always available. For instance, Fe^{3+} binds strongly to FA and HA (Milne et al. 2003), but there are very few detailed binding studies available in literature, making the model constants less certain. Very recently, the Fe^{3+} binding has been modelled for marine DOM (Hiemstra and Van Riemsdijk 2006b). For a verification of the applicability of a model approach to field samples it is also important to be able to measure the free ion concentration (Hiemstra and Van Riemsdijk 2006b).

The third problem in applying models based on purified materials to natural samples is that in natural samples several types of particles may be present that may interact with each other. The interactions between the colloidal particles may affect the ion binding to the individual particles and lead to deviation from the sum of ion adsorption when there are no interactions between the particles. More discussion on this topic will be given in Sect. 8.5 of this chapter. One also has to characterize the soil or sediment with respect to the amount of reversibly adsorbed ions per unit amount of soil. This amount is not necessarily equal to the total amount of an element, since the ions can also be present in the structure of minerals present in the soil, and these ions do not participate in the adsorptive binding.

8.4.5 Multi-Surface Model

Despite of the above-mentioned difficulties, attempts have been made to apply these ion adsorption models to understand and to predict ion speciation in environmental systems, given the present state of the art in this field. One of these attempts was made by Weng et al. (Weng et al. 2001a), who measured the concentration of a series of metal ions in equilibrium with the soil and compared these measurements with model predictions. In this study, three solid adsorbent phases were recognized in the soil, *i.e.* NOM, illite clay and iron (hydr)oxides. The site density of these surfaces was estimated. For the clay, the charge density is assumed to be equal to that of a pure illite, which is the dominant clay for the soils in this study. The site density of NOM was calculated from the difference of the measured CEC (cation exchange capacity) of the soil and the charge present on the clay (clay content times CEC of the clay). It was further assumed that the estimated charge of the organic matter is due to the charge of generic humic acid for the pH of the soil sample. The determined reactive humic acid content (31 %) is lower than the NOM content. For iron (hydr)oxides, the total amount of amorphous or crystalline iron (hydr)oxides was derived from respectively the oxalate extractable iron content and the difference between the DithioniteCitrateBicarbonate and oxalate extractable iron. The surface area of amorphous iron (hydr)oxides was assumed to be equal to that of HFO with 600 m^2/g, and that of the crystalline iron (hydr)oxides to be equal to goethite with 100 m^2/g. The competition of Al^{3+} ion to the binding was taken into account by assuming that its activity was controlled by gibbsite solubility. Competition with Fe^{3+} ion was not considered. The effect of the interaction between the adsorbent surfaces on the ion binding was ignored. Ion adsorption to illite clay, to NOM, to crystalline and amorphous iron (hydr)oxides was calculated with respectively the Donnan model, the NICA-Donnan model, the CD-MUSIC model and the Generalized Two Layer model (Dzombak and Morel 1990). The predicted metal ion activity of Cu^{2+}, Cd^{2+}, Zn^{2+}, Ni^{2+} and Pb^{2+} was compared to that measured with the Donnan membrane technique (DMT) (Temminghoff et al. 2000, Weng et al. 2001b). The results show that the agreement is good for Cu, Cd, Zn and Ni. For Pb, the model overestimated the metal activity. Examples of the results are given in Fig. 8.4 for Cu and Cd.

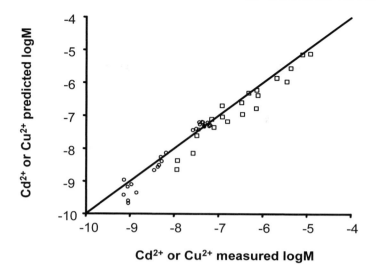

Fig. 8.4. Comparison of predicted (y-axis) and measured (x-axis) metal activities. Squared symbols are Cu^{2+} activity, round symbols are Cd^{2+} activity. Line is the 1 to 1 line between the predicted and measured values. (Weng et al. 2001a).

For the soils studied by Weng et al. (Weng et al. 2001a), the pH is in the acidic range. In another study by Schroder et al. (Schroder et al. 2005), a similar modelling approach has been adopted to predict metal ion speciation in river floodplain soils that have a pH range of 5-8. The modelling method is slightly different from that of Weng et al. In the work of Schroder et al., it was assumed that 50% of the solid soil organic matter is HA, the rest is inert. It was further assumed that the dissolved organic matter in the soil solution contains 40% FA, the rest being inert. The calculated total dissolved metal concentration was compared to those measured. The results show that for Cu there is a good agreement between the prediction and measurement. For Cd, the agreement is reasonable, but the model overestimated the dissolved Cd concentration by about 0.5 log-unit. For Zn and Pb, the concentration was overestimated by more than one log-unit. The authors improved the goodness of fit by introducing mineral precipitation for Zn and Pb. In another study following a similar modelling approach, Cances et al. (Cances et al. 2003) found good agreement between the free ion activity predicted and measured with the Donnan membrane technique for Cu, Cd, Zn and Pb for soils in the pH range of 4-6. In another study by Dijkstra et al. (Dijkstra et al. 2004b), a similar multi-surface model was used to simulate the leaching of heavy metals from contaminated soils over a wide range of pH (pH 0.4-12). Differing from the work of Weng et al. and Schroder et al., the site density of NOM was measured. The "reactive organic matter" was measured by measuring the content of HA and FA and it was found that the sum of HA and FA in the eight soil samples ranges from 25% to 67% of total NOM and on average 46%, which is similar to what Weng et al. has estimated from the measured soil CEC and measured clay content. The NICA-Donnan model, the Generalized Two Layer model (Dzombak and Morel 1990) and the Donnan model was used respectively to calculate ion binding to NOM, iron and aluminium (hydr)oxides and clay silicates. The results show that the model prediction of the leaching of Cu, Cd, Ni, Zn and Pb is in general adequate, and sometimes excellent. A similar approach has also been used by Lumsdon (Lumsdon 2004) to describe the solid-solution partitioning of Cd and Al in several soil profiles and good agreement between model prediction and data was found.

8.5 Adsorption of Natural Organic Matter to Minerals

8.5.1 Adsorption of NOM to Oxides

In the pH range of most natural environments, humic substances are negatively charged, and the oxide minerals are mostly positively charged. The opposite charge/potential carried by the NOM and oxide particles favours the interactions between the two. Humic substances adsorb strongly onto oxide minerals (Tipping 1981, Davis 1982, Murphy et al. 1992, Gu et al. 1994, Spark et al. 1997, Vermeer and Koopal 1998, Filius et al. 2000, Saito et al. 2004). Fig. 8.5 shows the adsorption envelope of Strichen FA on goethite (Filius et al. 2000). The adsorption of NOM to mineral surfaces may influence the solubility and mobility of the natural organic particles in the environment, and therefore the solubility and mobility of the elements bound to it. Organic matter can enhance (when dissolved) or retard (when in the solid matrix) the transport of the contaminants through the environment (Carter and Suffet 1983, McCarthy and Zachara 1989, Stevenson 1994, Benedetti et al. 1995, Jordan et al. 1997, Temminghoff et al. 1998). The humics present as dissolved organic matter (DOM) will in general play a direct and/or indirect role in the problem of transport of pollutants via mobile colloids. The direct role may be humics in the aqueous phase of the porous medium that will move with the water and can transport pollutants bound to it. It may have a strong indirect role when the mobile colloids are of inorganic nature, like clay or metal (hydr)oxides, because the humics will interact with these inorganic colloids in the aqueous phase affecting their mobility and their adsorption characteristics. The organic coating on the mineral surface will change the surface properties of the mineral and will influence the colloidal stability of the mineral particles. The solubility of NOM can be influenced by the source of the NOM, activity of micro-organism, solution chemistry and the composition of the solid phase. The solubility of the NOM can be controlled by the colloidal stability of the material or by adsorption to the solid matrix. Understanding the mechanisms that control the solubility of NOM is also crucial with respect to the carbon cycle in the environment.

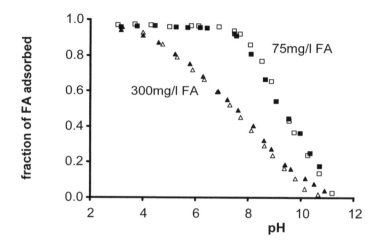

Fig. 8.5. Fractional adsorption of Strichen FA on goethite in $NaNO_3$ solution (Filius et al. 2000). Empty symbols: 0.015 M $NaNO_3$; filled: 0.10 M $NaNO_3$. Amount of goethite used is 5 g/l.

8.5.2 Effects of NOM Adsorption on Ion Binding

The interactions between NOM particles and minerals not only influence the solubility of NOM, but also affect the ion binding to both the NOM and oxides (Robertson and Leckie 1994, Zachara et al. 1994, Murphy and Zachara 1995, Vermeer et al. 1999, Christl and Kretzschmar 2001a, Weng et al. 2005). Most metal cations can bind to both humics and metal (hydr)oxides. The interactions between the humics and oxides influence the binding of metal cations to both particles by site competition, electrostatic effects and formation of new ternary surface complexes. The final outcome with respect to the metal ion binding to the complex is the net result of various interactions, and depends on the nature of the mineral surface and the humic, the type of metal ion and factors such as pH and ionic strength. The sum of metal ions bound to the mixture may deviate from the additive sum by assuming no particle interaction. For example, in a system containing Ca, FA and goethite, the amount of Ca bound is more than the additive sum at high pH but less than the additive sum at low pH (Fig. 8.6) (Weng et al. 2005).

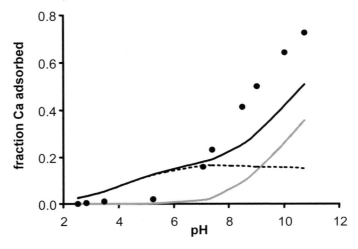

Fig. 8.6. Ca (1.0 mM) adsorption to the mixture of Strichen FA (150 mg/l) and goethite (6 g/l) in 0.1M NaNO$_3$ (Weng et al. 2005). Symbols are experimental data. Dotted line is Ca bound to adsorbed FA predicted using the NICA-Donnan model. Grey line is Ca bound to goethite predicted with the CD-MUSIC model. The solid line is the sum of Ca bound to FA and goethite predicted assuming linear additivity (Weng et al. 2005).

Note that the calculated amount of calcium bound to the FA that is bound to the oxide decreases at high pH (Fig. 8.6). This is not because the calcium binding to FA decreases at high pH, but because the amount of FA bound to the oxide decreases at high pH. Due to the complication that the humics interact strongly with the metal (hydr)oxide surface, direct application of either the CD-MUSIC model or the NICA-Donnan model to describe the adsorption of metal ions to a mixture of various particles leads to inaccurate predictions. For anions, which bind strongly to metal (hydr)oxides and in general not to the humics, the effects of the particle interactions will reduce the affinity of the anions for the coated oxide surface

(Grafe et al. 2001, Kreller et al. 2003, Hur and Schlautman 2004, Mukhopadhyay and Sanyal 2004, Borggaard et al. 2005). This competitive effect can be of both chemical and electrostatic nature. To be able to describe and predict the binding of anions to metal (hydr)oxides in the presence of humics, like for environmental samples, a model framework that can deal with the adsorption of both inorganic ions and organic particles in a consistent manner is required. The lack of such a model system is one of the major reasons that hinder the application of surface complexation models such as the CD-MUSIC model to field systems.

8.5.3 Challenges in Modelling Particle Adsorption

Particles have a large number of reactive ligands on each molecule. Therefore number of possible species that the particles may exhibit in both the solution and at the surface is huge. In practice, it is not possible to take all these possible species into account in the model. For particles in solution, the problem of the large number of possible species is "solved" by applying surface complexation models (SCM). Surface complexation models are often used to describe the reactions of the particle ligands with small ions. Most of the SCMs combine an ion-binding model, *e.g.* Langmuir, Langmuir-Freundlich and NICA with an electrostatic model, *e.g.* diffuse double layer, basic Stern and Donnan. These SCMs calculate the fraction (θ_i) of the ligands on the particles that is bound with a certain ion i and the total amount of ion bound to the particles (N_i). For the competitive Langmuir model this is expressed as:

$$N_i = N_{max} \theta_i = N_{max} \frac{K_i C_i}{1 + \sum_i (K_i C_i)}, \qquad (8.4)$$

where θ_i is the fraction of the ligands that are complexed with component i, K_i is the intrinsic affinity constant for the reaction between the ligand and component i, and C_i (mol/l) is the local concentration of i, N_{max} and N_i are respectively the total number of ligands on the particle and the ligands that are bound to i.

The SCMs calculate the statistical average distribution of the ligands among various possible states, *e.g.* dissociated, protonated, *etc.* For instance, using the Langmuir equation as given in Eq. 8.4, one can calculate the fraction of ligands that is protonated (θ_H). For ligands with equal energy (homogeneous), this means that the probability to find a ligand that is protoned equals θ_H. If the total number of ligands per particle is N_{max}, and all ligands are equal, the probability to find a fully protonated particle is therefore θ_H^{Nmax}. A similar approach can be applied to describe the chemical state of the adsorbed particles. Surface complexation models that have been used to describe ion binding to particles in solution can also be applied to particles adsorbed to a surface. In this case, the local ion concentration at the adsorption surface has to be used in the model to calculate the adsorption of small ions. If chemical binding between the particle ligands and sites on the macro surface is considered, the surface sites can be treated in the SCMs similar to binding of small ions to the particle. In this way, the state of the particle in the adsorbed state can be calculated and the average distribution of the ligands of the adsorbed particles over all its possible states can be derived. The advantage of using the SCMs to describe particles is that it doesn't invoke the definition of explicit species and still can take all possibilities of speciation into account by calculating the average.

The macroscopic ion complexation models are convenient to describe the average state of the particles in one phase. However, to calculate the amount of particles that are adsorbed to a charged surface, one cannot simply follow the method discussed above for the calculation of the adsorption of small molecules. Due to the interactions between the particles and the surface, the average chemical state of the particles will change upon adsorption, as well the average charge of the particles. Because of these effects, the adsorption of the particle cannot be calculated using the classical thermodynamic approach for small ions. The change of the chemical state of the particle is a measure of the free energy change of the particles and can be used to calculate the adsorption free energy and the adsorption isotherm for variable charged particles to charged surfaces (Chan and Mitchell 1983, Stigter and Dill 1989, Overbeek 1990, McCormack et al. 1995, Biesheuvel 2004a, Biesheuvel and Stuart 2004b, Biesheuvel et al. 2005).

8.5.4 Adsorption Energy of Variable Charge Particles

The concept of the free energy of variable charge particles or the free energy of the electrical double layer has been illustrated previously in the literature (Chan and Mitchell 1983, Stigter and Dill 1989, Overbeek 1990). The change of the free energy of the variable charge particles due to ion binding (specifically and non-specifically) can be calculated from:

$$dF = -\sum N_i d\mu_i \qquad (8.5)$$

where dF (J/mol) is the change of the (Helmholtz) free energy per mol particles, N_i (mol/mol) is the (excess) amount of ion i bound per mol particles and $d\mu_i$ is the change of the chemical potential of ion i. For the specifically bound ions, the amount adsorbed (N_i) can be calculated using one of the ion-binding models such as the Langmuir, Langmuir-Freundlich and NICA model. When combined with electrostatic models, the chemical binding calculated with these ion-binding models includes the electrostatic effect on the chemical binding. For non-specifically bound ions, the amount adsorbed (N_i) can be calculated from the particle charge and the electrostatic model chosen.

The (electro)chemical potential $(\mu_p$, J/mol) of the particles can be written as:

$$\mu_p = \mu_p^0 + \Delta F_{sp} + \Delta F_{nsp} + RT\ln C_p \tag{8.6}$$

where μ_p^0 (J/mol) is the standard chemical potential of the particles, which can be defined as the chemical potential of one mol particle at an arbitrarily chosen reference state (often the neutral uncharged state is used as reference state), ΔF_{sp} and ΔF_{nsp} (J/mol) are the change of the free energy of the particles from the chosen reference state due to the binding or release of respectively specifically and non-specifically bound ions, C_p (mol/l) is the total concentration of the particles in the phase of concern, R and T are respectively the gas constant and absolute temperature. The change of the free energy of the particles (ΔF) can be calculated by taking the integral of Eq. 8.5.

At equilibrium, the (electro)chemical potential (μ_p) of the particles in the solution phase and in the adsorption phase should be equal, which leads to:

$$C_{p,ads} = C_{p,sol}e^{\left(\frac{\Delta F_{sp,sol}-\Delta F_{sp,ads}}{RT}\right)}e^{\left(\frac{\Delta F_{nsp,sol}-\Delta F_{nsp,ads}}{RT}\right)} = C_{p,sol}K_{p,sp}K_{p,nsp}, \tag{8.7}$$

where subscripts 'sol' refers to the solution phase and 'ads' the adsorption phase to which the particles adsorb, K_p is the adsorption affinity of the particles, which is divided into a specific part $(K_{p,sp})$ and a non-specific part $(K_{p,nsp})$.

The change of the free energy of the particles can be calculated by integrating Eq. 8.5. For instance, if we choose the neutral fully protonated particles as the reference, when protons are the only specifically bound ions in the system, the change of the free energy of the particles due to the release of protons can be calculated as:

$$\Delta F_{sp} = \int_{\infty}^{\mu_H} dF_{sp} = \int_{\infty}^{\mu_H} (N_{max} - N_H) d\mu_H$$

$$= RTN_{max} \int_{\infty}^{C_H} (N_{max} - N_H) d\ln C_H ,$$

(8.8)

where N_{max} (mol/mol) is the total number of ligands per particle, C_H is the local concentration of protons. The limits for the integral in Eq. 8.8 indicate that the integral is calculated from infinitely high proton concentration to the actual 'local' proton concentration. At an infinitely high proton concentration, the particles will be fully protonated, which is the chosen reference state. The concept given in Eq. 8.8 can be further illustrated with the help of a "master curve" approach (De Wit et al. 1990, De Wit et al. 1993a). Fig. 8.7 shows a charge/pH curve of a particle calculated with the Langmuir model (solid line), in which the number of protons bound per particle (N_H) is plotted as a function of the local pH. Because the local proton concentration equals the proton activity in the bulk solution times the Boltzmann factor of the local phase, the local pH equals: $pH_{sol} - \log B$. When protons are the only ions that form complexes with the particles in the system, the change of protonation of the particle is determined only by the local proton activity. This type of curve has been called the "master curve" (De Wit et al. 1990, De Wit et al. 1993a). The horizontal scale in Fig. 8.7 is proportional (times 2.3RT) to the change of the chemical potential of the protons. Integration of the number of protons adsorbed or released per particle (N_H or $N_{max}-N_H$) over the chemical potential of the proton, results in the free energy change of the particle due to protonation/deprotonation.

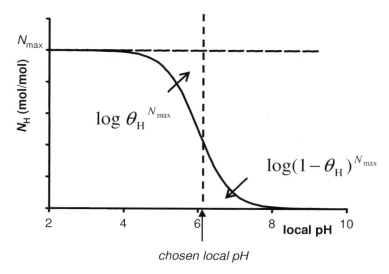

chosen local pH

Fig. 8.7 The master curve of the protonation/deprotonation of a particle described with the Langmuir model (solid line). The indicated integrated area of this curve over the chemical potential of protons equals the free energy change of the particle due to protonation/deprotonation.

Based on the above approach, expressions for the free energy change have been derived for the NICA model (Weng et al. 2006):

$$\Delta F_{sp} = RT \frac{N_{max}}{n_H p} \left\{ \ln(1 - \sum \theta_i) + \ln(\tilde{K}_H C_H)^{n_H p} \right\} \tag{8.9}$$

and the Donnan model for a system that contains a symmetric 1:1 electrolyte ions (Weng et al. 2006):

$$\Delta F_{nsp} = -C_{salt} V_{D,m} RT (B_D + \frac{1}{B_D} - 2), \tag{8.10}$$

where C_{salt} is the concentration of the electrolyte ions (monovalent) in the solution, $V_{D,m}$ is the Donnan volume of one mol particles (l/mol), B_D is the Boltzmann factor for the Donnan phase. From the change of the free energy (Eqs. 8.9 and 8.10), the adsorption affinity of the particles can be expressed as:

$$K_{p,sp} = \left(\frac{1 - \sum \theta_{i,sol}}{1 - \sum \theta_{i,ads}} \right)^{\frac{N_{max}}{n_H p}} \left(\frac{B_{sol}}{B_{ads}} \right)^{N_{max}} \qquad (8.11)$$

$$K_{p,nsp} = e^{\frac{C_{salt} V_{D,m} f}{RT} \left(B_{D,ads} + \frac{1}{B_{D,ads}} - B_{D,sol} - \frac{1}{B_{D,sol}} \right)} \qquad (8.12)$$

The beauty of the NICA equation is that it can be integrated and results in a new very elegant analytical equation that describes the essence of the binding of variable charge nano-particles to charged surfaces (Eq. 8.11).

8.5.5 LCD Model Concept

The concept of the Ligand and Charge Distribution (LCD) model has been proposed to describe the adsorption of macromolecules or particles, especially humic substances, to metal (hydr)oxides (Filius et al. 2000, Filius et al. 2003, Weng et al. 2005). The goal of the development of the LCD model is to have a model structure that can calculate the adsorption of both small ions and particles simultaneously in a consistent way and that can make use of the details already present in the advanced surface complexation models for oxides and humics. The LCD model derives its name from the fact that the statistical distribution of the functional groups, and the charge of the bound particles over the interface are part of the model output.

The equilibrium condition for the phase distribution of the variable charge particles is a crucial part for models dealing with particle adsorption. In the previous version of the LCD model, the adsorption of the particles is calculated using the overall reaction formalism (Filius et al. 2000, Filius et al. 2003, Weng et al. 2005). In this formalism, the activity of fully dissociated humic or fulvic molecule is used, which is however, not correctly defined in the previous versions of LCD. The general reaction method used in the previous version of the LCD model leads to different predictions as to the phase distribution of the particles than those predicted using the free energy approach as given in Eq. 8.5.

The availability of the new approach to calculate the equilibrium distribution of variable charge particles makes it possible to improve the LCD model. In doing so, the general structure of the model framework will remain the same as in the previous version. However, the general reaction formalism used in the previous version to calculate the equilibrium phase distribution of particles needs to be replaced by the expressions derived based on the free energy concept. In the LCD, the NICA model is used to calculate the average chemical state of humics in the solution phase and adsorption phase. The essence of the use of the NICA concept in the LCD model is that not only small ions like protons but also reactive surface sites of the oxides compete for interaction with the reactive ligands of the adsorbed humic particles. The electrostatic potentials at the position that the ligands are located are the potentials calculated in the CD-MUSIC model at the corresponding planes. The LCD model has been implemented in the computer code ORCHESTRA (Meeussen 2003) and the calculation is carried out numerically in an iterative way. The improved LCD model is still under development and its ability to describe humics adsorption in colloid-colloid systems and in ion-colloid-colloid systems still needs to be tested.

8.5.6 Vision for Future Development

Steady progress has been made in the last two decades with respect to the development of physical chemical models that can handle competitive binding of ions to important reactive surfaces like humics and metal (hydr)oxides. The present models for humics have much less molecular detail than the models for metal (hydr)oxides. The humic ion binding models are more of a statistical, thermodynamic nature, which seems appropriate because of the ill-defined nature of the mixture of humic substances that are present in nature. Application of the NICA-Donnan model to various practical situations shows great promise for its applicability to environmental systems. Our understanding of the ion binding to metal (hydr)oxides on the molecular level has increased considerably in the last two decades. The CD-MUSIC model tries to incorporate as much of the information that is available on the molecular level, without compromising its usefulness for efficient speciation calculations. Ion binding to metal (hydr)oxides in natural systems is very complex due to the interaction with humics and a large series of inorganic cations and anions. At present the oxide models have not yet been developed and tested far enough to be able to deal with the reality of the complexity of environmental systems.

The large variety of different types of (electrostatic) models that have been developed has led some authors to go in the direction of non-electrostatic models for ion binding (Bradbury and Baeyens 2005). From a thermodynamic perspective one can create large families of models that are all thermodynamically correct. However, the capability for the prediction of the effect of the interaction between ions as present in nature may differ considerably between various model approaches. For environmental application it is very important that the model that is used properly describes the multi component interaction, including the interaction with humic substances. Since the interactions are to a large extent of electrostatic nature, it is an advantage to use an electrostatic approach that takes the structure of the interface and the adsorbed species into account as good as possible.

The recent progress in some fundamental aspects of humic binding to oxides as described above, is a big step forward in the direction of the applicability of these models to environmental systems. There is urgent need for a further development and testing of the LCD model. More refined techniques, than the crude ones used at present, are needed to estimate the reactive area of metal (hydr)oxides as present in environmental samples. Further developments are also necessary with respect to measuring and modelling ion binding to the edge faces of phyllo-silicate clay minerals for conditions that are relevant in nature. It is to be expected that the applicability of this type of approach to environmental systems will rapidly increase in the coming years. A further step forward is to consider the structure of soil aggregates and to combine the fundamental approach highlighted in this chapter with dynamic reactive ion transport, since slow kinetics are often very relevant for the behaviour of ions like phosphate. The reactive transport approach can also be used to better understand the processes that take place in the rhizosphere. The models have the potential to better understand major effects that are relevant at the field scale or larger scale. Natural processes are often of biogeochemical nature. For understanding of the bioavailability of nutrients and contaminants, process level understanding is important. The approach sketched above should also find an increasing use in risk assessment studies that are the basis for the development of legal limits as set by governments with respect to the quality of the environment.

Acknowledgements

Part of the work presented in this chapter was funded by the EU project, FUNMIG (516514 (F16W-2004).

References

Ali MA, Dzombak DA (1996) Comparative sorption of simple organic acids and sulfate on goethite. Environmental Science and Technology 30 (4):1061-1071

Allison JD, Brown DS et al. (1991) MINTEQA2/PRODEFA2-A geochemical assessment model for environmental systems: Version 3.0 User's Manual Environmental Protection Agency

Anderson PR, Christensen TH (1988) Distribution Coefficients of Cd, Co, Ni, and Zn in Soils. Journal of Soil Science 39 (1):15-22

Antelo J, Avena M et al. (2005) Effects of pH and ionic strength on the adsorption of phosphate and arsenate at the goethite-water interface. Journal of Colloid and Interface Science 285 (2):476-486

Arai Y, Sparks DL (2001) ATR-FTIR spectroscopic investigation on phosphate adsorption mechanisms at the ferrihydrite-water interface. Journal of Colloid and Interface Science 241 (2):317-326

Atkinson RJ, Posner AM et al. (1967) Adsorption of potential-determining ions at the ferric oxide-aqueous electrolyte interface. Journal of Physical Chemistry 71 550-558

Balnois E, Wilkinson KJ et al. (1999) Atomic force microscopy of humic substances: Effects of pH and ionic strength. Environmental Science & Technology 33 (21):3911-3917

Bargar JR, Fuller CC et al. (2005) Nanoparticulate bacteriogenic manganese oxides: Environmental reactivity and structural chemistry. Geochimica Et Cosmochimica Acta 69 (10):A461-A461

Bargar JR, Reitmeyer R et al. (1999) Spectroscopic confirmation of uranium(VI)-carbonato adsorption complexes on hematite. Environmental Science & Technology 33 (14):2481-2484

Bargar JR, Reitmeyer R et al. (2000) Characterization of U(VI)-carbonato ternary complexes on hematite: EXAFS and electrophoretic mobility measurements. Geochimica Et Cosmochimica Acta 64 (16):2737-2749

Barrow NJ, Bowden JW Reactions with Variable-Charge Soils (1987) Martinus Nijhof Publishers Dordrecht

Benedetti MF, Milne CJ et al. (1995) Metal-Ion binding to humic substances - application of the nonideal competitive adsorption model. Environmental Science & Technology 29 (2):446-457

Benedetti MF, Van Riemsdijk WH et al. (1996) Humic substances considered as a heterogeneous donnan gel phase. Environmental Science & Technology 30 (6):1805-1813

Bickmore BR, Tadanier CJ et al. (2004) Bond-Valence methods for pK(a) prediction: critical reanalysis and a new approach. Geochimica Et Cosmochimica Acta 68 (9):2025-2042

Biesheuvel PM (2004a) Electrostatic free energy of interacting ionizable double layers. Journal of Colloid and Interface Science 275 (2):514-522

Biesheuvel PM, Stuart MAC (2004b) Electrostatic free energy of weakly charged macromolecules in solution and intermacromolecular complexes consisting of oppositely charged polymers. Langmuir 20 (7):2785-2791

Biesheuvel PM, Van der Veen M et al. (2005) A modified Poisson-Boltzmann model including charge regulation for the adsorption of ionizable polyelectrolytes to charged interfaces, applied to lysozyme adsorption on silica. Journal of Physical Chemistry B 109 (9):4172-4180

Boily JF, Persson P et al. (2000) Benzenecarboxylate surface complexation at the goethite (alpha-FeOOH)/water interface: II. Linking IR spectroscopic observations to mechanistic surface complexation models for phthalate, trimellitate, and pyromellitate. Geochimica Et Cosmochimica Acta 64 (20):3453-3470

Borggaard OK, Raben-Lange B et al. (2005) Influence of humic substances on phosphate adsorption by aluminium and iron oxides. Geoderma 127 (3-4):270-279

Borkovec M (1997) Origin of 1-pK and 2-pK models for ionizable water-solid interfaces. Langmuir 13 (10):2608-2613

Bourikas K, Hiemstra T et al. (2001a) Adsorption of molybdate monomers and polymers on titania with a multisite approach. Journal of Physical Chemistry B 105 (12):2393-2403

Bourikas K, Hiemstra T et al. (2001b) Adsorption of molybdate monomers and polymers on titania with a multisite approach. Journal of Physical Chemistry B 105 (12):2393-2403

Bowden JW, Posner AM et al. (1977) Ionic Adsorption on Variable Charge Mineral Surfaces - Theoretical-Charge Development and Titration Curves. Australian Journal of Soil Research 15 (2):121-136

Bradbury MH, Baeyens B (2005) Modelling the sorption of Mn(II), Co(II), Ni(II), Zn(II), Cd(II), Eu(III), Am(III), Sn(IV), Th(IV), Np(V) and U(VI) on montmorillonite: Linear free energy relationships and estimates of surface binding constants for some selected heavy metals and actinides. Geochimica Et Cosmochimica Acta 69 (4):875-892

Buffle J Complexation Reactions in Aquatic Systems, Ellis Horwood (1988)

Buffle J, Wilkinson KJ et al. (1998) A generalized description of aquatic colloidal interactions: The three-colloidal component approach. Environmental Science & Technology 32 (19):2887-2899

Cances B, Ponthieu M et al. (2003) Metal ions speciation in a soil and its solution: experimental data and model results. Geoderma 113 (3-4):341-355

Carter CW, Suffet IH Fate of Chemicals in the Environment. Compartmental and Multimedia Models for Predictions (1983) American Chemical Society; Washington, D.C.; USA 215-229

Catalano JG, Park C et al. (2006) Termination and water adsorption at the alpha-Al2O3 (012) - Aqueous solution interface. Langmuir 22 (10):4668-4673

Chan DYC, Mitchell DJ (1983) The Free-Energy of an Electrical Double-Layer. Journal of Colloid and Interface Science 95 (1):193-197

Christl I, Kretzschmar R (1999) Competitive sorption of copper and lead at the oxide-water interface: Implications for surface site density. Geochimica Et Cosmochimica Acta 63 (19-20):2929-2938

Christl I, Kretzschmar R (2001a) Interaction of copper and fulvic acid at the hematite-water interface. Geochimica Et Cosmochimica Acta 65 (20):3435-3442

Christl I, Kretzschmar R (2001b) Relating ion binding by fulvic and humic acids to chemical composition and molecular size. 1. Proton binding. Environmental Science & Technology 35 (12):2505-2511

Christl I, Milne CJ et al. (2001c) Relating ion binding by fulvic and humic acids to chemical composition and molecular size. 2. Metal binding. Environmental Science & Technology 35 (12):2512-2517

Collins CR, Ragnarsdottir KV et al. (1999) Effect of inorganic and organic ligands on the mechanism of cadmium sorption to goethite. Geochimica Et Cosmochimica Acta 63 (19-20):2989-3002

Combes JM, Chisholmbrause CJ et al. (1992) EXAFS Spectroscopic Study of Neptunium(V) Sorption at the Alpha-FeOOH Water Interface. Environmental Science & Technology 26 (2):376-382

Connor PA, Dobson KD et al. (1999) Infrared spectroscopy of the TiO2/aqueous solution interface. Langmuir 15 (7):2402-2408

Davis JA (1982) Adsorption of natural dissolved organic matter at the oxide/water interface. Geochimica Et Cosmochimica Acta 46 2381-2393

Davis JA, James RO et al. (1978) Surface Ionization and Complexation at Oxide-Water Interface .1. Computation of Electrical Double-Layer Properties in Simple Electrolytes. Journal of Colloid and Interface Science 63 (3):480-499

De Wit JCM, Van Riemsdijk WH et al. (1993a) Proton Binding to Humic Substances .1. Electrostatic Effects. Environmental Science & Technology 27 (10):2005-2014

De Wit JCM, Van Riemsdijk WH et al. (1993b) Proton Binding to Humic Substances .2. Chemical Heterogeneity and Adsorption Models. Environmental Science & Technology 27 (10):2015-2022

De Wit JCM, Van Riemsdijk WH et al. (1990) Analysis of Ion Binding on Humic Substances and the Determination of Intrinsic Affinity Distributions. Analytica Chimica Acta 232 (1):189-207

Degryse F, Broos K et al. (2003) Soil solution concentration of Cd and Zn can be predicted with a CaCl2 soil extract. European Journal of Soil Science 54 (1):149-157

Dent AJ, Ramsay JDF et al. (1992) An EXAFS Study of Uranyl-Ion in Solution and Sorbed onto Silica and Montmorillonite Clay Colloids. Journal of Colloid and Interface Science 150 (1):45-60

Dijkstra JJ, Meeussen JCL et al. (2004a) Leaching of heavy metals from contaminated soils: An experimental and modeling study. Environmental Science & Technology 38 (16):4390-4395

Dijkstra JJ, Meeussen JCL et al. (2004b) Leaching of heavy metals from contaminated soils: An experimental and modeling study. Environmental Science & Technology 38 (16):4390-4395

Dudal Y, Gerard F (2004) Accounting for natural organic matter in aqueous chemical equilibrium models: a review of the theories and applications. Earth-Science Reviews 66 (3-4):199-216

Duval JFL, Wilkinson KJ et al. (2005) Humic substances are soft and permeable: Evidence from their electrophoretic mobilities. Environmental Science & Technology 39 (17):6435-6445

Dzombak DA, Morel FMM Surface Complexation Modeling: Hydrous Ferric Oxide, Wiley, New York (1990)

Elzinga EJ, Van Grinsven JJM et al. (1999) General purpose Freundlich isotherms for cadmium, copper and zinc in soils. European Journal of Soil Science 50 (1):139-149

Felmy AR, Rustad JR (1998) Molecular statics calculations of proton binding to goethite surfaces: Thermodynamic modeling of the surface charging and protonation of goethite in aqueous solution. Geochimica Et Cosmochimica Acta 62 (1):25-31

Filius JD, Hiemstra T et al. (1997a) Adsorption of small weak organic acids on goethite: Modeling of mechanisms. Journal of Colloid and Interface Science 195 (2):368-380

Filius JD, Hiemstra T et al. (1997b) Adsorption of small weak organic acids on goethite: Modeling of mechanisms. Journal of Colloid and Interface Science 195 (2):368-380

Filius JD, Lumsdon DG et al. (2000) Adsorption of fulvic acid on goethite. Geochimica Et Cosmochimica Acta 64 (1):51-60

Filius JD, Meeussen JCL et al. (2001) Modeling the binding of benzenecarboxylates by goethite: The ligand and charge distribution model. Journal of Colloid and Interface Science 244 (1):31-42

Filius JD, Meeussen JCL et al. (2003) Modeling the binding of fulvic acid by goethite: the speciation of adsorbed FA molecules. Geochimica Et Cosmochimica Acta 67 (8):1463-1474

Geelhoed JS, Hiemstra T et al. (1998) Competitive interaction between phosphate and citrate on goethite. Environmental Science & Technology 32 (14):2119-2123

Goldberg S, Sposito G (1984) A Chemical-Model of Phosphate Adsorption by Soils .1. Reference Oxide Minerals. Soil Science Society of America Journal 48 (4):772-778

Grafe M, Eick MJ et al. (2001) Adsorption of arsenate (V) and arsenite (III) on goethite in the presence and absence of dissolved organic carbon. Soil Science Society of America Journal 65 (6):1680-1687

Gu BH, Schmitt J et al. (1994) Adsorption and Desorption of Natural Organic-Matter on Iron- Oxide - Mechanisms and Models. Environmental Science & Technology 28 (1):38-46

Gustafsson JP (2001) Modelling competitive anion adsorption on oxide minerals and an allophane-containing soil. European Journal of Soil Science 52 (4):639-653

Gustafsson JP (2003) Modelling molybdate and tungstate adsorption to ferrihydrite. Chemical Geology 200 (1-2):105-115

Hayes KF, Leckie JO (1986) Mechanism of Lead-Ion Adsorption at the Goethite-Water Interface. Acs Symposium Series 323 114-141

Hayes KF, Roe AL et al. (1987) Insitu X-Ray Absorption Study of Surface Complexes - Selenium Oxyanions on Alpha-FeOOH. Science 238 (4828):783-786

Hayes MHB, MacCarthy P et al. Humic Substances II. In Search of Structure, John Wiley & Sons Ltd (1989)

Hesterberg D, Bril J et al. (1993) Thermodynamic Modeling of Zinc, Cadmium, and Copper Solubilities in a Manured, Acidic Loamy-Sand Topsoil. Journal of Environmental Quality 22 (4):681-688

Hiemstra T, De Wit JCM et al. (1989b) Multisite proton adsorption modeling at the solid-solution interface of (hydr)oxides - a new approach. 2. application to various important (hydr)oxides. Journal of Colloid and Interface Science 133 (1):105-117

Hiemstra T, Rahnemaie R et al. (2004a) Surface complexation of carbonate on goethite: IR spectroscopy, structure and charge distribution. Journal of Colloid and Interface Science 278 (2):282-290

Hiemstra T, Rahnemaie R et al. (2004b) Interpretation of surface species from spectroscopy and CD modelling. Geochimica Et Cosmochimica Acta 68 (11):A159-A159

Hiemstra T, Van Riemsdijk WH (1991) Physical-chemical interpretation of primary charging behavior of metal (hydr)oxides. Colloids and Surfaces 59 7-25

Hiemstra T, Van Riemsdijk WH (1996) A surface structural approach to ion adsorption: The charge distribution (CD) model. Journal of Colloid and Interface Science 179 (2):488-508

Hiemstra T, Van Riemsdijk WH (1999a) Surface structural ion adsorption modeling of competitive binding of oxyanions by metal (hydr)oxides. Journal of Colloid and Interface Science 210 (1):182-193

Hiemstra T, Van Riemsdijk WH (1999b) Effect of different crystal faces on experimental interaction force and aggregation of hematite. Langmuir 15 (23):8045-8051

Hiemstra T, Van Riemsdijk WH (2000) Fluoride adsorption on goethite in relation to different types of surface sites. Journal of Colloid and Interface Science 225 (1):94-104

Hiemstra T, Van Riemsdijk WH Encyclopedia of Surface and Colloid Science (2002) Marcel Dekker New York 3773-3799

Hiemstra T, Van Riemsdijk WH (2004) Interfacial distribution of charge: Relationship between micro- and macroscopic ion adsorption phenomena. Abstracts of Papers of the American Chemical Society 227 U1215-U1215

Hiemstra T, Van Riemsdijk WH (2006b) Biogeochemical speciation of Fe in ocean water. Marine Chemistry 102 (3-4):181-197

Hiemstra T, Van Riemsdijk WH et al. (1989a) Multisite proton adsorption modeling at the solid-solution interface of (hydr)oxides - a new Approach. 1. Model description and evaluation of intrinsic reaction constants. Journal of Colloid and Interface Science 133 (1):91-104

Hiemstra T, Van Riemsdijk WH et al. (1987) Proton adsorption mechanism at the gibbsite and aluminium oxide solid/solution interface. Netherlands Journal of Agricultural Science 35 (3):281-293

Hiemstra T, Venema P et al. (1996) Intrinsic proton affinity of reactive surface groups of metal (hydr)oxides: The bond valence principle. Journal of Colloid and Interface Science 184 (2):680-692

Hiemstra T, Yong H et al. (1999) Interfacial charging phenomena of aluminum (hydr)oxides. Langmuir 15 (18):5942-5955

Hingston FJ, Posner AM et al. (1971) Competitive adsorption of negatively charged ligands on oxide surfaces. Discussions of the Faraday Society (52):334-342

Hug SJ (1997) In situ Fourier transform infrared measurements of sulfate adsorption on hematite in aqueous solutions. Journal of Colloid and Interface Science 188 (2):415-422

Hur J, Schlautman MA (2004) Effects of pH and phosphate on the adsorptive fractionation of purified Aldrich humic acid on kaolinite and hematite. Journal of Colloid and Interface Science 277 (2):264-270

Impellitteri CA, Saxe JK et al. (2003) Predicting the bioavailability of copper and zinc in soils: Modeling the partitioning of potentially bioavailable copper and zinc from soil solid to soil solution. Environmental Toxicology and Chemistry 22 (6):1380-1386

Janssen RPT, Peijnenburg W et al. (1997) Equilibrium partitioning of heavy metals in Dutch field soils .1. Relationship between metal partition coefficients and soil characteristics. Environmental Toxicology and Chemistry 16 (12):2470-2478

Jing CY, Meng XG et al. (2005) Surface complexation of organic arsenic on nanocrystalline titanium oxide. Journal of Colloid and Interface Science 290 (1):14-21

Jolivet JP, Froidefond C et al. (2004) Size tailoring of oxide nanoparticles by precipitation in aqueous medium. A semi-quantitative modelling. Journal of Materials Chemistry 14 (21):3281-3288

Jordan RN, Yonge DR et al. (1997) Enhanced mobility of Pb in the presence of dissolved natural organic matter. Journal of Contaminant Hydrology 29 (1):59-80

Karltun E (1998) Modelling SO42- surface complexation on variable charge minerals. II. Competition between SO42-, oxalate and fulvate. European Journal of Soil Science 49 (1):113-120

Keizer MG, Van Riemsdijk WH (1994) ECOSAT: Equilibrium Calculation of Speciation and Transport Agricultural University of Wageningen

Kim CS, Rytuba JJ et al. (2004) EXAFS study of mercury(II) sorption to Fe- and Al-(hydr)oxides I. Effects of pH. Journal of Colloid and Interface Science 271 (1):1-15

Kinniburgh DG, Milne CJ et al. (1996) Metal ion binding by humic acid: Application of the NICA-Donnan model. Environmental Science & Technology 30 (5):1687-1698

Kinniburgh DG, Van Riemsdijk WH et al. (1999) Ion binding to natural organic matter: competition, heterogeneity, stoichiometry and thermodynamic consistency. Colloids and Surfaces a-Physicochemical and Engineering Aspects 151 (1-2):147-166

Koopal LK, Saito T et al. (2005) Ion binding to natural organic matter: General considerations and the NICA-Donnan model. Colloids and Surfaces a-Physicochemical and Engineering Aspects 265 (1-3):40-54

Koopal LK, Van Riemsdijk WH et al. (1994) Analytical isotherm equations for multicomponent adsorption to heterogeneous surfaces. Journal of Colloid and Interface Science 166 (1):51-60

Kreller DI, Gibson G et al. (2003) Competitive adsorption of phosphate and carboxylate with natural organic matter on hydrous iron oxides as investigated by chemical force microscopy. Colloids and Surfaces a-Physicochemical and Engineering Aspects 212 (2-3):249-264

Lovgren L, Sjoberg S et al. (1990) Acid-Base Reactions and Al(III) Complexation at the Surface of Goethite. Geochimica Et Cosmochimica Acta 54 (5):1301-1306

Lumsdon DG (2004) Partitioning of organic carbon, aluminium and cadmium between solid and solution in soils: application of a mineral-humic particle additivity model. European Journal of Soil Science 55 (2):271-285

Lutzenkirchen J (1998) Comparison of 1-pK and 2-pK versions of surface complexation theory by the goodness of fit in describing surface charge data of (hydr)oxides. Environmental Science & Technology 32 (20):3149-3154

Machesky ML, Wesolowski DJ et al. (2001) On the temperature dependence of intrinsic surface protonation equilibrium constants: An extension of the revised MUSIC model. Journal of Colloid and Interface Science 239 (2):314-327

Manceau A, Charlet L (1994) The Mechanism of Selenate Adsorption on Goethite and Hydrous Ferric-Oxide. Journal of Colloid and Interface Science 168 (1):87-93

Manceau A, Gorshkov AI et al. (1992) Structural Chemistry of Mn, Fe, Co, and Ni in Manganese Hydrous Oxides .1. Information from XANES Spectroscopy. American Mineralogist 77 (11-12):1133-1143

McCarthy JF, Zachara JM (1989) Subsurface Transport of Contaminants - Mobile Colloids in the Subsurface Environment May Alter the Transport of Contaminants. Environmental Science & Technology 23 (5):496-502

McCormack D, Carnie SL et al. (1995) Calculations of Electric Double-Layer Force and Interaction Free-Energy between Dissimilar Surfaces. Journal of Colloid and Interface Science 169 (1):177-196

Meeussen JCL (2003) ORCHESTRA: An object-oriented framework for implementing chemical equilibrium models. Environmental Science & Technology 37 (6):1175-1182

Milne CJ, Kinniburgh DG et al. (2001) Generic NICA-Donnan model parameters for proton binding by humic substances. Environmental Science & Technology 35 (10):2049-2059

Milne CJ, Kinniburgh DG et al. (2003) Generic NICA-Donnan model parameters for metal-ion binding by humic substances. Environmental Science & Technology 37 (5):958-971

Mouvet C, Bourg ACM (1983) Speciation (Including Adsorbed Species) of Copper, Lead, Nickel and Zinc in the Meuse River - Observed Results Compared to Values Calculated with a Chemical-Equilibrium Computer-Program. Water Research 17 (6):641-649

Mukhopadhyay D, Sanyal SK (2004) Complexation and release isotherm of arsenic in arsenic-humic/fulvic equilibrium study. Australian Journal of Soil Research 42 (7):815-824

Murphy EM, Zachara JM (1995) The Role of Sorbed Humic Substances on the Distribution of Organic and Inorganic Contaminants in Groundwater. Geoderma 67 (1-2):103-124

Murphy EM, Zachara JM et al. (1992) The Sorption of Humic Acids to Mineral Surfaces and Their Role in Contaminant Binding. Science of the Total Environment 118 413-423

Nabavi M, Spalla O et al. (1993) Surface-Chemistry of Nanometric Ceria Particles in Aqueous Dispersions. Journal of Colloid and Interface Science 160 (2):459-471

Oakley SM, Nelson PO et al. (1981) Model of Trace-Metal Partitioning in Marine-Sediments. Environmental Science & Technology 15 (4):474-480

Ostergren JD, Bargar JR et al. (1999) Combined EXAFS and FTIR investigation of sulfate and carbonate effects on Pb(II) sorption to goethite (alpha-FeOOH). Journal of Synchrotron Radiation 6 645-647

Overbeek JTG (1990) The Role of Energy and Entropy in the Electrical Double-Layer. Colloids and Surfaces 51 61-75

Parfitt RL, Russell JD et al. (1976) Confirmation of Surface-Structures of Goethite (Alpha-FeOOH) and Phosphated Goethite by Infrared Spectroscopy. Journal of the Chemical Society-Faraday Transactions I 72 1082-1087

Parkhurst DL, Appelo CAJ (2002) User's guide to PHREEQC (version 2), a computer program for speciation, batch-reaction, one-dimensional transport, and inverse geochemical calculations US Geological Survey

Parkman RH, Charnock JM et al. (1999) Reactions of copper and cadmium ions in aqueous solution with goethite, lepidocrocite, mackinawite, and pyrite. American Mineralogist 84 (3):407-419

Pashley RM, Israelachvili JN (1984) Molecular layering of water in thin films between mica surfaces and its relation to hydration forces. Journal of Colloid and Interface Science 101 (2):511-523

Peak D, Ford RG et al. (1999) An in situ ATR-FTIR investigation of sulfate bonding mechanisms on goethite. Journal of Colloid and Interface Science 218 (1):289-299

Pinheiro JP, Mota AM et al. (1999) Lead and calcium binding to fulvic acids: Salt effect and competition. Environmental Science & Technology 33 (19):3398-3404

Pinheiro JP, Mota AM et al. (2000) Effect of aluminum competition on lead and cadmium binding to humic acids at variable ionic strength. Environmental Science & Technology 34 (24):5137-5143

Rahnemaie R, Hiemstra T et al. (2006a) A new structural approach for outersphere complexation tracing the location of electrolyte ions. Journal of Colloid and Interface Science 293 312-321

Rahnemaie R, Hiemstra T et al. (2006b) Inner- and outersphere complexation of ions at the goethite-solution interface. Journal of Colloid and Interface Science 297 379-388

Randall SR, Sherman DM et al. (1999) The mechanism of cadmium surface complexation on iron oxyhydroxide minerals. Geochimica Et Cosmochimica Acta 63 (19-20):2971-2987

Redden G, Bargar J et al. (2001) Citrate enhanced uranyl adsorption on goethite: An EXAFS analysis. Journal of Colloid and Interface Science 244 (1):211-219

Ridley MK, Machesky ML et al. (2002) Potentiometric studies of the rutile-water interface: hydrogen-electrode concentration-cell versus glass-electrode titrations. Colloids and Surfaces a-Physicochemical and Engineering Aspects 204 (1-3):295-308

Ridley MK, Machesky ML et al. (2004) Modeling the surface complexation of calcium at the rutile-water interface to 250 degrees C. Geochimica Et Cosmochimica Acta 68 (2):239-251

Rietra RPJJ, Hiemstra T et al. (1999) The relationship between molecular structure and ion adsorption on variable charge minerals. Geochimica Et Cosmochimica Acta 63 (19-20):3009-3015

Rietra RPJJ, Hiemstra T et al. (2001a) Comparison of selenate and sulfate adsorption on goethite. Journal of Colloid and Interface Science 240 (2):384-390

Rietra RPJJ, Hiemstra T et al. (2001b) Interaction between calcium and phosphate adsorption on goethite. Environmental Science & Technology 35 (16):3369-3374

Rochester CH, Topham SA (1979) Infrared Study of Surface Hydroxyl-Groups on Goethite. Journal of the Chemical Society-Faraday Transactions I 75 591-602

Rose J, Flank AM et al. (1997) Nucleation and growth mechanisms of Fe oxyhydroxide in the presence of PO4 ions .2. P K-edge EXAFS study. Langmuir 13 (6):1827-1834

Rudzinski W, Charmas R et al. (1999) Searching for thermodynamic relations in ion adsorption at oxide/electrolyte interfaces studied by using the 2-pK protonation model. Langmuir 15 (25):8553-8557

Rusch U, Borkovec M et al. (1997) Interpretation of competitive adsorption isotherms in terms of affinity distributions. Journal of Colloid and Interface Science 191 247-255

Sahai N, Carroll SA et al. (2000) X-ray absorption spectroscopy of strontium(II) coordination - II. Sorption and precipitation at kaolinite, amorphous silica, and goethite surfaces. Journal of Colloid and Interface Science 222 (2):198-212

Sahai N, Sverjensky DA (1997) Solvation and electrostatic model for specific electrolyte adsorption. Geochimica Et Cosmochimica Acta 61 (14):2827-2848

Saito T, Koopal LK et al. (2004) Adsorption of humic acid on goethite: Isotherms, charge adjustments, and potential profiles. Langmuir 20 (3):689-700

Sauve S, Manna S et al. (2003) Solid-solution partitioning of Cd, Cu, Ni, Pb, and Zn in the organic horizons of a forest soil. Environmental Science & Technology 37 (22):5191-5196

Sauve S, Norvell WA et al. (2000) Speciation and complexation of cadmium in extracted soil solutions. Environmental Science & Technology 34 (2):291-296

Schindler PW, Kamber HR (1968) Acidity of silanol groups. Preliminary study. Helvetica Chimica Acta 51 (7):1781-&

Schroder TJ, Hiemstra T et al. (2005) Modeling of the solid-solution partitioning of heavy metals and arsenic in embanked flood plain soils of the rivers Rhine and Meuse. Environmental Science & Technology 39 (18):7176-7184

Schudel M, Behrens SH et al. (1997) Absolute aggregation rate constants of hematite particles in aqueous suspensions: A comparison of two different surface morphologies. Journal of Colloid and Interface Science 196 (2):241-253

Sherman DM, Randall SR (2003) Surface complexation of arsenate(V) to iron(III) (hydr)oxides: Structural mechanism from ab initio molecular geometries and EXAFS spectroscopy. Geochimica Et Cosmochimica Acta 67 (22):4223-4230

Sigg L, Stumm W (1981) The Interaction of Anions and Weak Acids with the Hydrous Goethite (Alpha-FeOOH) Surface. Colloids and Surfaces 2 (2):101-117

Sips R (1948) On the Structure of a Catalyst Surface. Journal of Chemical Physics 16 (5):490-495

Spadini L, Manceau A et al. (1994) Structure and Stability of Cd2+ Surface Complexes on Ferric Oxides .1. Results from EXAFS Spectroscopy. Journal of Colloid and Interface Science 168 (1):73-86

Spadini L, Schindler PW et al. (2003) Hydrous ferric oxide: evaluation of Cd-HFO surface complexation models combining Cd-K EXAFS data, potentiometric titration results, and surface site structures identified from mineralogical knowledge. Journal of Colloid and Interface Science 266 (1):1-18

Spark KM, Wells JD et al. (1997) Characteristics of the sorption of humic acid by soil minerals. Australian Journal of Soil Research 35 (1):103-112

Stachowicz M, Hiemstra T et al. (2006) Surface speciation of As(III) and As(V) adsorption in relation to charge distribution. Journal of Colloid and Interface Science 302 62-75

Stevenson FJ Humus Chemistry: Genesis, Composition, Reactions, John Wiley and Sons; New York; USA (1994)

Stigter D, Dill KA (1989) Free-Energy of Electrical Double-Layers - Entropy of Adsorbed Ions and the Binding Polynomial. Journal of Physical Chemistry 93 (18):6737-6743

Strawn DG, Sparks DL (2000) Effects of soil organic matter on the kinetics and mechanisms of Pb(II) sorption and desorption in soil. Soil Science Society of America Journal 64 (1):144-156

Sverjensky DA (1994) Zero-Point-of-Charge Prediction from Crystal-Chemistry and Solvation Theory. Geochimica Et Cosmochimica Acta 58 (14):3123-3129

Sverjensky DA (2005) Prediction of surface charge on oxides in salt solutions: Revisions for 1 : 1 (M+L-) electrolytes. Geochimica Et Cosmochimica Acta 69 (2):225-257

Tadanier CJ, Eick MJ (2002) Formulating the charge-distribution multisite surface complexation model using FITEQL. Soil Science Society of America Journal 66 (5):1505-1517

Tejedor & Tejedor MI, Anderson MA (1990) Protonation of Phosphate on the Surface of Goethite as Studied by CIR-FTIR and Electrophoretic Mobility. Langmuir 6 (3):602-611

Temminghoff EJM, Plette ACC et al. (2000) Determination of the chemical speciation of trace metals in aqueous systems by the Wageningen Donnan Membrane Technique. Analytica Chimica Acta 417 (2):149-157

Temminghoff EJM, Van der Zee SEATM et al. (1995) Speciation and calcium competition effects on cadmium sorption by sandy soil at various pHs. European Journal of Soil Science 46 (4):649-655

Temminghoff EJM, Van der Zee SEATM et al. (1997) Copper mobility in a copper-contaminated sandy soil as affected by pH and solid and dissolved organic matter. Environmental Science & Technology 31 (4):1109-1115

Temminghoff EJM, Van der Zee SEATM et al. (1998) Effects of dissolved organic matter on the mobility of copper in a contaminated sandy soil. European Journal of Soil Science 49 (4):617-628

Temminghoff EJM, Van der Zee SEATM et al. (1994) The Influence of Ph on the Desorption and Speciation of Copper in a Sandy Soil. Soil Science 158 (6):398-408

Thurman EM Organic Geochemistry of Natural Waters, Nijhoff Junk, Dordrecht (1985)

Tipping E (1981) The adsorption of aquatic humic substances by iron oxides. Geochimica Et Cosmochimica Acta 45 191-199

Tipping E (1994) WHAM - a chemical equilibrium model and computer code for waters, sediments, and soils incorporating a discrete site/electrostatic model of ion-binding by humic substances. Computers and Geosciences 20 (6):973-1023

Tipping E (1998) Humic ion-binding model VI: An improved description of the interactions of protons and metal ions with humic substances. Aquatic Geochemistry 4 (1):3-48

Tipping E Cation Binding by Humic Substances, Cambridge University Press (2002)

Tipping E, Hurley MA (1992) A unifying model of cation binding by humic substances. Geochimica Et Cosmochimica Acta 56 (10):3627-3641

Tournassat C, Ferrage E et al. (2004) The titration of clay minerals II. Structure-based model and implications for clay reactivity. Journal of Colloid and Interface Science 273 (1):234-246

Venema P, Hiemstra T et al. (1996a) Comparison of different site binding models for cation sorption: Description of pH dependency, salt dependency, and cation-proton exchange. Journal of Colloid and Interface Science 181 (1):45-59

Venema P, Hiemstra T et al. (1996b) Multisite adsorption of cadmium on goethite. Journal of Colloid and Interface Science 183 (2):515-527

Venema P, Hiemstra T et al. (1998) Intrinsic proton affinity of reactive surface groups of metal (hydr)oxides: Application to iron (hydr)oxides. Journal of Colloid and Interface Science 198 (2):282-295

Vermeer AWP, Koopal LK (1998) Adsorption of humic acids to mineral particles. 2. Polydispersity effects with polyelectrolyte adsorption. Langmuir 14 (15):4210-4216

Vermeer AWP, McCulloch JK et al. (1999) Metal ion adsorption to complexes of humic acid and metal oxides: Deviations from the additivity rule. Environmental Science & Technology 33 (21):3892-3897

Villalobos M, Leckie JO (2001) Surface complexation modeling and FTIR study of carbonate adsorption to goethite. Journal of Colloid and Interface Science 235 (1):15-32

Waychunas GA, Rea BA et al. (1993) Surface-Chemistry of Ferrihydrite .1. EXAFS Studies of the Geometry of Coprecipitated and Adsorbed Arsenate. Geochimica Et Cosmochimica Acta 57 (10):2251-2269

Weerasooriya R, Tobschall HJ et al. (2003) On the mechanistic modeling of As(III) adsorption on gibbsite. Chemosphere 51 (9):1001-1013

Weerasooriya R, Tobschall HJ et al. (2004) Macroscopic and vibration spectroscopic evidence for specific bonding of arsenate on gibbsite. Chemosphere 55 (9):1259-1270

Weerasooriya R, Wijesekara H et al. (2002) Surface complexation modeling of cadmium adsorption on gibbsite. Colloids and Surfaces a-Physicochemical and Engineering Aspects 207 (1-3):13-24

Weng LP, Koopal LK et al. (2005) Interactions of calcium and fulvic acid at the goethite-water interface. Geochimica Et Cosmochimica Acta 69 (2):325-339

Weng LP, Temminghoff EJM et al. (2002) Complexation with dissolved organic matter and solubility control of heavy metals in a sandy soil. Environmental Science & Technology 36 (22):4804-4810

Weng LP, Temminghoff EJM et al. (2001a) Contribution of individual sorhents to the control of heavy metal activity in sandy soil. Environmental Science & Technology 35 (22):4436-4443

Weng LP, Temminghoff EJM et al. (2001b) Determination of the free ion concentration of trace metals in soil solution using a soil column Donnan membrane technique. European Journal of Soil Science 52 (4):629-637

Weng LP, Van Riemsdijk WH et al. (2006) Adsorption Free Energy of Variable-Charge Nanoparticles to a Charged Surface in Relation to the Change of the Average Chemical State of the Particles. Langmuir 22 389-397

Wesolowski DJ, Machesky ML et al. (2000) Magnetite surface charge studies to 290 degrees C from in situ pH titrations. Chemical Geology 167 (1-2):193-229

Westall J, Hohl H (1980) Comparison of Electrostatic Models for the Oxide-Solution Interface. Advances in Colloid and Interface Science 12 (4):265-294

Wijnja H, Schulthess CP (1999) ATR-FTIR and DRIFT spectroscopy of carbonate species at the aged gamma-Al2O3/water interface. Spectrochimica Acta Part a-Molecular and Biomolecular Spectroscopy 55 (4):861-872

Yates DE The Structure of the Oxide/Aqueous Electrolyte Interface., University of Melbourne (1975)

Yates DE, Levine S et al. (1974) Site-Binding Model of Electrical Double-Layer at Oxide-Water Interface. Journal of the Chemical Society-Faraday Transactions I 70 1807-1818

Zachara JM, Resch CT et al. (1994) Influence of Humic Substances on Co2+ Sorption by a Subsurface Mineral Separate and Its Mineralogic Components. Geochimica Et Cosmochimica Acta 58 (2):553-566

Zhang Z, Fenter P et al. (2004) Ion adsorption at the rutile-water interface: Linking molecular and macroscopic properties. Langmuir 20 (12):4954-4969

9 Release of Contaminants from Bottom Ashes - Colloid Facilitated Transport and Colloid Trace Analysis by Means of Laser-Induced Breakdown Detection (LIBD)

Rainer Köster[1], Tobias Wagner[1], Markus Delay[2], Fritz H. Frimmel[2]

[1] Institut für Technische Chemie, Bereich Wasser- und Geotechnologie, Forschungszentrum Karlsruhe GmbH, Postfach 3640, 76021 Karlsruhe, Germany
[2] Engler-Bunte-Institut, Chair of Water Chemistry, Universität Karlsruhe (TH), Engler-Bunte-Ring 1, 76131 Karlsruhe, Germany

9.1 Introduction

In Germany about 11 million t/a of municipal solid waste is incinerated giving 3 million t/a of ashes (slags). The assessment of the ashes and their possible usage get increasing attention worldwide. Several leaching tests have been developed. However, most of them consider the total concentrations of soluble species only.

A comprehensive and realistically important investigation of the geochemical long-term behavior of residues has to include colloidal particles in their transport function for contaminants (Tushar and Khilar 2006, Ryan and Elimelech 1996, Loveland et al. 2003). This transport is well-known from natural aquatic systems where the occurrence of colloids is ubiquitous (Stumm and Morgan 1996). Therefore, the colloidal particles are important for the distribution of pollutants in soil and aquifers as well. The function of the colloids is strongly influenced by their surface charge which is surface area related significantly higher than in bulk material with larger particles (McCarthy and Zachara 1989, McDowell-Boyer et al. 1986).

The object of this book chapter is to report on
1. the basic material behavior in column experiments,
2. the colloidal heavy metal release and

3.   the testing of a new, coupled column system with highly sensitive La-
     ser-Induced Breakdown Detection (LIBD) to characterize the released
     colloids in detail.

The investigated bottom ashes are multi-component matrices with a pH-
value of about 11 and higher. There are some toxic elements (e.g. Cu, Zn,
Pb) that are enriched up to two orders of magnitude compared to concentrations found in the lithosphere. The elemental composition of ashes is
similar to the one of the earth crust (Chandler et al. 1997). Mineralogical
studies have shown that molten materials like thermodynamically instable
glassy phases of different elemental composition are present in addition to
minerals such as calcite, haematite or silicates like melilite and calcium
silicate hydrate phases (CSH phases) (Zevenbergen et al. 1994). Due to
this initial situation the long-term mobility of the trace metals is one of the
main concerns with regard to the use of bottom ashes as a secondary raw
material. Many investigations have been performed regarding the solubility of trace elements/compounds in highly alkaline conditions that are
typical for bottom ash leachates. Higher solubility of the trace elements or
trace element compounds can only be expected in the acidic pH-value
range. Due to the alkaline pH-value of the leachate, one of the main parameters influencing a potential long-term trace metal release is the acid
neutralizing capacity (Johnson et al. 1995).

Additional aspects of mobilization via colloids, especially against the
backdrop of thermodynamically unstable phases in the ashes, have been
neglected most often; the main reason for this is that common particle
analysis technology has been not sensitive enough for characterizing colloids in their natural environment. Thus, colloid bound transport was investigated in this work, and the Laser-Induced Breakdown Detection was
introduced as a highly sensitive tool for characterizing particles in liquids.
In the past, LIBD has proven to be a suitable tool for monitoring particle
removal during drinking water purification, storage and distribution, for
characterizing different filter systems, and also for the investigation of colloid facilitated transport of contaminants. The positive results have in-
spired experiments to couple the LIBD directly to a column leaching unit
in order to quantify the particle release from waste materials (e.g. bottom
ashes from municipal waste incinerators or construction sites) into the
aquatic environment.

9.2 Methodical Background of the LIBD

9.2.1 Theoretical Aspects

One of the major difficulties when dealing with colloids is that many analysis techniques are invasive. This means that their application has an influence on the properties of the usually complex original system. A rather new and promising method to overcome part of these problems is the LIBD. The method is capable of characterizing particles in liquid media down to around 10 nm in size (even lower with special optical setups), and with unprecedented resolution in the ng/L (ppt) concentration range.

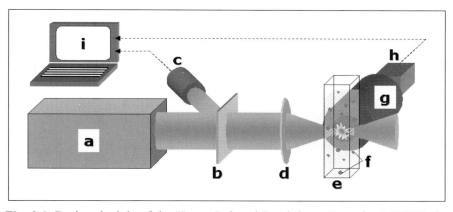

Fig. 9.1. Basic principle of the "Laser-Induced Breakdown Detection" (LIBD) for characterizing colloids regarding size, number density and mass concentration. a) pulsed laser (20 Hz); b) beam splitter; c) energy detector; d) lens; e) sample cell; f) colloids; g) microscope; h) camera; i) computer.

The basic idea behind the method is to focus a pulsed laser beam into the sample cell (Fig. 9.1), and to make use of the fact that for igniting a plasma (dielectric breakdown of matter) on solid particles (colloids), a significantly lower energy density is required than for a plasma generation in particle free liquids like water (Radziemski and Cremers 1989). This means that the laser energy can be adjusted so that every time the pulsed laser beam hits a particle in the sample, this very particle will be evaporated, accompanied by a plasma flash and an acoustic shock wave. On the other hand, if the laser beam "hits" the dispersive agent only (commonly water), this will cause no noticeable effect. It is easily understandable that the more particles in the sample, and the bigger those particles are, the more often such plasma events can be detected; that is to say the higher is the so-called breakdown probability is (number of hits per total number of laser pulses) (Bundschuh 1999, Kitamori et al. 1988).

Mathematically, this translates to

$$W_{\mathrm{Bd}} = 1 - \left(1 - c_{\mathrm{p}} V_{\mathrm{P}}\right)^{\frac{V_{\mathrm{F,eff}}(\mathrm{P})}{V_{\mathrm{P}}}} \qquad (9.1)$$

W_{Bd}:   Breakdown probability
c_{P}:   Particle concentration in $1/\mathrm{m}^3$
V_{P}:   Particle volume in m^3
$V_{\mathrm{F,eff}}(\mathrm{P})$:   Effective focal volume of particle P in m^3,

where $V_{\mathrm{F,eff}}(\mathrm{P})$ is

$$V_{\mathrm{F,eff}}(P) = \frac{4}{3}\pi^2 \cdot n(\lambda) \cdot \frac{r_{\mathrm{e}}(z'=0)^4}{\lambda} \sqrt{\left(\frac{A_{\mathrm{P}}}{C} - 1\right) \cdot ln\left(\frac{A_{\mathrm{P}}}{C}\right)} \qquad (9.2)$$

$n(\lambda)$:   Refractive index at wavelength λ
λ: Laser wavelength in m
$r_{\mathrm{e}}(z'=0)$:   Beam waist in m
A_{P}:   Particle area in m^2
C: Minimal particle area able to generate plasma in m^2

A detailed derivation of the above formulas is described in the literature (Bundschuh et al. 2005, Scherbaum et al. 1996).

It has been found that the spatial distribution of such plasma events along the laser beam axis is a function of the particle size only, and not of the particle concentration (Bundschuh et al. 2001); see Fig. 9.2. This means that in the course of a measurement a few thousand plasma events have to be recorded. The function varies with the nature of the suspended colloids. Organic compounds and biocolloids usually show a similar behavior than the polystyrene spheres often used for standard calibration. The behavior of inorganic compounds may be different, and although it seems that the deviation is far less than an order of magnitude, further investigations are presently being carried out.

In an unknown sample first the mean particle size has to be determined, and subsequently the particle concentration and number density can be calculated from the breakdown probability (Scherbaum et al. 1996). This is illustrated in Fig. 9.3.

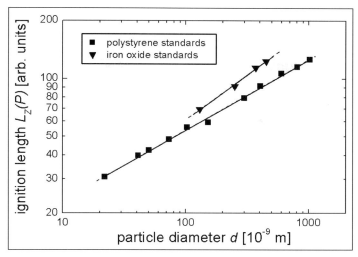

Fig. 9.2. Ignition length (spatial distribution of plasma events) along laser beam axis as a function of the colloid size. $L_Z(P)$ simply is the ±3 fold standard deviation of the coordinates of the single plasma events of one measurement.

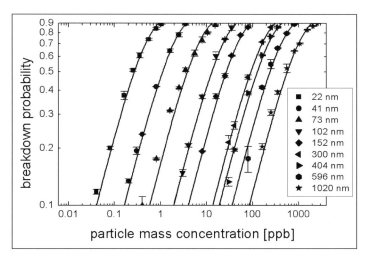

Fig. 9.3. Breakdown probability as a function of colloid concentration. The lines represent the calculated breakdown probability according to the model by Scherbaum (1996).

Compared to laser light scattering and obscuration the LIBD shows a better sensitivity by several orders of magnitude (Fig. 9.4), especially for particles < 100 nm; it is also capable of detecting bacteria and viruses (Bundschuh et al. 2005, Wagner 2005). This predestines the technology for characterization of colloids in natural aquatic systems where they are the predominant species and mostly found in concentrations low enough to make sample pretreatment (e.g. preconcentration) necessary for conventional analysis techniques (Kim et al. 1992). The LIBD usually does not require sample preparation, except for dilution if the breakdown probability is too high ("overmodulation of detector"). Also, the method is commonly known to be non-invasive; the number of colloids evaporated due to plasma ignition in the course of a measurement can be neglected. Single measurements typically take a few minutes, the sample volume required is about 2 mL, and flow-trough cells for *online* measurements also can be used without problems. Since the dynamic range of the LIBD for a given particle size fraction is about 1 to 2 orders of magnitude, the laser pulse energy should be adjusted prior to online measurements in order to avoid the necessity of online dilution. However, a new measurement mode (patent pending) will enhance the dynamic range noticeably in the near future.

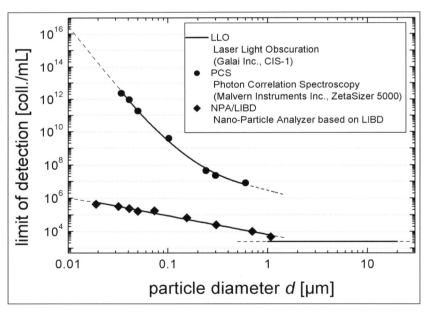

Fig. 9.4. Comparison of detection limits of different particle analysis techniques.

9.2.2 Features of the LIBD

At present, the instrumentation cannot distinguish between organic, inorganic or biological particles as it does not reveal their chemical composition. In the future, this may be partly achieved by using spectrometers to analyze the plasma emission of evaporated particles.

For samples containing particles of different size the LIBD returns a mean particle diameter – compared to light scattering methods, it is number weighted rather than intensity weighted. To obtain a particle size distribution, pre-fractionation is required, e.g. by *online*-coupling with field-flow fractionation; this has already been demonstrated successfully (Thang et al. 2000, Wagner 2005).

Particles larger than 1 μm do not disturb the measurement, but determining their size becomes increasingly inaccurate. The LIBD works best in the colloidal size range, i.e. from 10 - 1000 nm.

The material dependence of the breakdown process causes measurement errors for unknown samples, in unfortunate cases this can be a factor of more than 2. For standard calibration polystyrene latex spheres are used; as a consequence, measurement results are polystyrene equivalents. Still, there is no other method available to date that allows measurement of natural aquatic colloids (without pre-treatment) with better precision. In most such cases, the colloids are too small and/or present in too low concentrations.

Bearing all this in mind, the LIBD is best suited for examination of changes in particle populations (both known and unknown samples), and for characterization of colloids of known nature. Furthermore, the instrumentation can be used as highly sensitive particle detector, both in combination with fractionation techniques or as a threshold detector that triggers an alarm once a certain particle threshold in a liquid phase is reached.

9.3 Materials and Methods

9.3.1 Colloid Facilitated Heavy Metal Mobilization

The overall procedure for the investigations is given in Fig. 9.5. Based on basic material characterization of the slags, the colloidal effects were investigated. The slags had a relatively high salt content, high pH-value and a high acid neutralization capacity. Two different approaches with batch and column experiments for colloid mobilization were chosen (see below). In accordance with German legislation for material reuse, the slags were stored for three months before investigation (storage under natural weathering conditions).

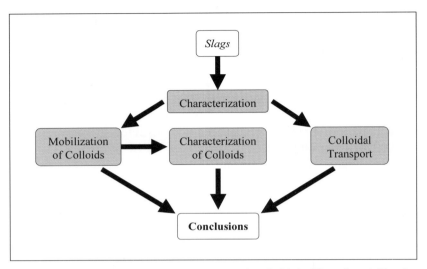

Fig. 9.5. Overall procedure for investigation of colloid facilitated mobilization of heavy metals.

The set-up of the first type of column experiments is given in Fig. 9.6 (second type with online coupling to LIBD is depicted in Fig. 9.7). The acrylic glass columns had an inner diameter of 20 mm and a length of 106 mm. They were packed wet with about 30 g of ash. The flow rate of the aqueous eluent (alternately deionized water and salt solutions) was 1.1 ml/min, supplied by a HPLC pump (Knauer K-501) connected in series to a degasser. pH-value and electrical conductivity were permanently measured online. The elemental analysis of Na, Ca, Si, Al and Fe in the eluent were monitored over the entire time of the experiments by means of a sequential optical emission spectrometry with inductively coupled plasma (ICP-OES) (Jobin Yvon, JY 38 S).

The soluble species from the ash column were removed by a preceding elution step with 0.5 M NaCl solution. Then the colloid mobilization was initiated by consecutive changes between a NaCl containing eluent and deionized water in equidistant time intervals.

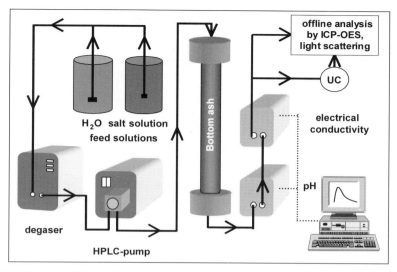

Fig. 9.6. Set-up of the column experiments to investigate colloid mobilization out of bottom ashes. Sieve fractions of small particles are eluted alternately with 0.5 mol/L NaCl solution and pure water (UC: ultra centrifugation).

Differentiation between the colloidal and dissolved element fraction was achieved by centrifugation of the samples (Sorvall Pro 80 ultracentrifuge, T865 rotor). After 3 hours of centrifugation the calculated sample cut-off was approximately 5 nm (maximal centrifugal force: 254,000 g, assumed particle density: 2.5 g/cm³). For additional verification of the existence of colloidal matter the scattering intensities of some samples were examined by means of photon correlation spectroscopy (Malvern Instruments, Zetasizer 5000).

9.3.2 Coupling of a Column Leaching Unit with LIBD

Two standard materials of the German Federal Office for Material Research and Testing ("Bundesanstalt für Materialforschung und -prüfung" (BAM)) have been examined; described here are column leaching experiments with a bottom ash from municipal waste incinerators. The most important material properties are listed in Table 9.1.

Table 9.1. Most important material properties of the examined standard bottom ash from municipal waste incinerators (data according to Berger et al. 2004).

Parameter	Dimension	bottom ash
residual humidity	weight %	3.7
pH-value (20 °C) (H_2O) [a]	-	11.1
grain size distribution		
6.3 – 4 mm	weight %	-
4 – 2 mm	weight %	19.5
2 – 0.63 mm	weight %	26.8
0.63 – 0.2 mm	weight %	31.0
0.2 – 0.063 mm	weight %	17.6
< 0.063 mm	weight %	5.0
element content in solid [b]		
Cr	mg/kg dry mass	67
Cu	mg/kg dry mass	1657

[a] Eluate with water/solid ratio 2:1 L/kg
[b] Determination after dissolution with nitrohydrochloric acid.

The packed columns were eluted at a constant flow rate of 1.0 mL/min (HPLC pump Knauer K-501) with ultra-pure water (Sartorius 611 UF; pH-value of the eluent adjusted to 5) over a period of several hours, then the elution was stopped for a certain time. After this time the experiment was continued by switching on the pump again. The break in the elution process simulates the period between two rain showers.

Electrical conductivity and pH-value were measured online (Electrical conductivity Meter LF 538 with TetraCon 325 electrode and ProfiLab pH 597 with SenTix 41 electrode, both WTW GmbH), colloids were determined *online* by means of LIBD in combination with a flow-trough cell. The eluent was collected in 10 mL fractions. The selected fractions were analyzed with ICP-OES in order to reveal the elemental composition (without differentiation between particles and liquid); of particular interest were heavy metals. Furthermore, the fractions were examined by standard LIBD (*offline* mode) right after the elution experiments in order to obtain absolute values for particle size and concentration (more exactly speaking, the results are polystyrene equivalents as discussed earlier). Fig. 9.7 shows a schematic illustration of the experimental setup.

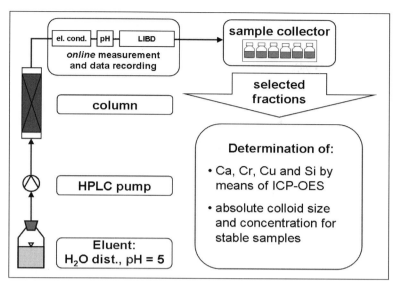

Fig. 9.7. Schematic setup of the column leaching experiments with online coupling to LIBD in order to investigate the mobilization of heavy metals and colloids out of bottom ashes from municipal waste incinerators.

9.4 Results

9.4.1 Colloidal Mobilization of Heavy Metals from Bottom Ashes

From batch experiments, a threshold concentration for the mobilization of colloids was found to be about 0.55 mmol/L calcium in the solute. Above this value, no colloid release occurred, but below, a strong increase occurred (particles detected in solution); it was strongly dependent on the calcium concentration in solution.

In the column experiments, the big change of the electrolyte concentration in solution leads to a readjustment of the equilibrium as a result of the competition between protons and sodium ions for the surface groups of the matrix. Corresponding to the loss or gain of protons in solution, a shift of the pH-value occurs upwards and downwards, respectively (SURF means surface in reaction equations below).

- Change of pH-value due to change in electrolyte composition:

  NaCl $\Rightarrow$ H$_2$O:

  SURF-O$^-$Na$^+$ + H$_2$O $\rightleftharpoons$ SURF-OH + Na$^+$ + **OH$^-$**

  H$_2$O $\Rightarrow$ NaCl:

  SURF-OH + Na$^+$ $\rightleftharpoons$ SURF-O$^-$Na$^+$ + **H$^+$**

- Calcium-sodium-equilibrium:

  NaCl $\Rightarrow$ H$_2$O:

  SURF-O$^-$Na$^+$ + Ca^{2+} $\rightleftharpoons$ SURF-O$^-$Ca^{2+} + Na$^+$

  H$_2$O $\Rightarrow$ NaCl:

  SURF-O$^-$Ca^{2+} + Na$^+$ $\rightleftharpoons$ SURF-O$^-$Na$^+$ + Ca^{2+}

After these shifts, the pH-value readjusts to values between 9.9 and 10.2, depending on the elution period, see Fig. 9.8. In the elution period with deionized water silicon and also aluminum behave very similar regarding development of the pH value. Their concentrations rise or fall, depending on the trend of the pH-value. As illustrated, nearly all measured iron was dispersed in colloidal form.

With both elution methods, the characterization of the leached colloids yielded a composition of calcium, silicon, aluminum, magnesium and iron as major elements. It turned out that significant amounts of the ecologically undesirable heavy metals copper, lead, zinc and chromium were bound to the colloidal matrix. Typical colloid bound heavy metal concentrations are in the order of mg/g slag in the leachate, thus in some cases exceeding the limits defined by German authorities (LAGA 1994); also see Fig. 9.9. Crystalline phases could not be assigned to the colloids.

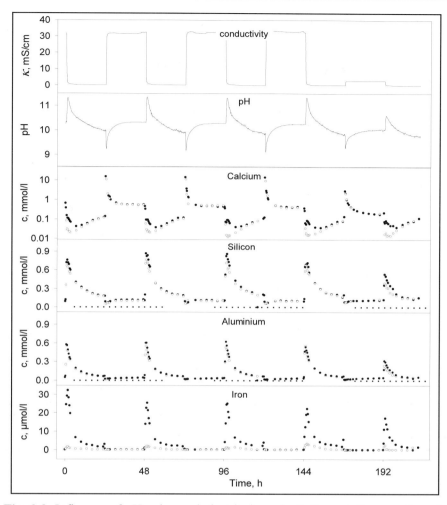

Fig. 9.8. Influence of pH value and electrical conductivity variations on the colloid mobilization in column experiments (closed symbols: before ultra centrifugation, open: after ultra centrifugation; chart as depicted in (Ferstl 2002)).

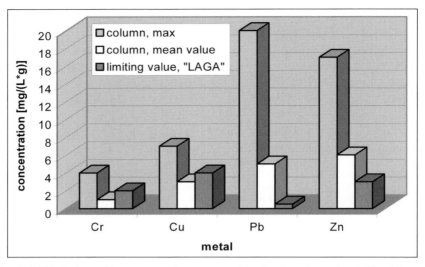

Fig. 9.9. Variation of heavy metal concentrations in the eluent (mobilized by colloids) per gram of slag.

In the column experiments two effects were obviously responsible for the colloid mobilization:

On the one hand, due to the moderate solubility of the calcium phases of slags, a decrease of colloid release was obtained with an increase of the calcium concentration of the solution. On the other hand, a dissolution of silicon phases occurred, so additional colloidal fragments were provided. For the latter effect amorphous and pH sensitive CSH phases are thought to be responsible. Further investigations with bottom ashes aged for five years in a lysimeter showed much lower colloid concentrations in the effluent; thus amorphous CSH phases have consolidated over time.

The profile of the iron concentration in the dispersion has a comparable shape to silicon. The peaks show less tailing which is attributed to the rising calcium concentration in solution which leads to destabilization of the colloids (electrical double layer becomes smaller). As illustrated in Fig. 9.8, nearly all measured iron was dispersed in colloidal form. At high ionic strengths, the iron concentration was always below the detection limit. Generally, if the electrolyte concentration was high, no difference in concentration was observed before and after centrifugation for all monitored elements. These results agree with photon correlation spectroscopy (PCS) measurements, where the signal intensity was just around the detection limit. Therefore, iron is a suitable and much better indicator for the existence of colloids than any other element (Ferstl 2002).

9.4.2 Online Detection of Particle Release with LIBD

Fairly high breakdown probabilities are recorded right after the start of the elution (overmodulation of detector) indicating the presence of high amounts of particles and colloids from the wash out (Fig. 9.10). In this elution stage, the electrical conductivity is at its maximum when compared to the further course of the experiment; it is attributed to readily soluble ions (solution and desorption processes); pH-value is also at its maximum – this is mainly due to dissolution of calcium hydroxide.

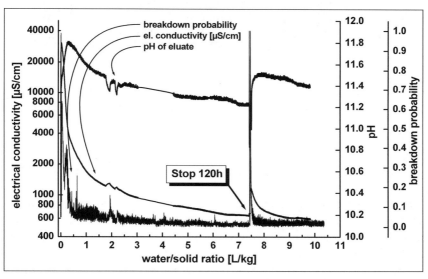

Fig. 9.10. Development of breakdown probability (particle signal), electrical conductivity and pH-value during the elution of bottom ashes from municipal waste incinerators. In the course of this experiment the flow was interrupted once for 120 hours.

The breakdown probability as well as electrical conductivity and pH-value decrease after a while. During the saturation of the column with the eluent the bottom ash is instantly not wetted thoroughly. This leads to subsequent fluctuations of the measurement signals (mainly seen in the particle signal). Furthermore, the decreasing ionic strength widens the electrical double layer of the colloids which opposes aggregation (Stumm and Morgan 1996) and in effect leads to secondary peaks in the breakdown signal caused by smaller particles.

Stopping the elution for a certain time (column remains saturated with eluent) and subsequent continuation of the experiment shows again a sharp rise of the breakdown probability. How long and how intense this temporary increase in particle mobilization is depends on the time of the flow interruption; the longer the stagnancy period, the bigger and longer the observed effect. Fig. 9.11 shows details, and it is interesting to note that for interruptions shorter than a certain threshold time (with this experimental setup about 2 hours) no significant increase in the particle signal is observed. Thus, the observed rises after flow interruptions longer than the threshold time are no artefacts caused by switching on the pump leading to a short and sudden increase of pressure.

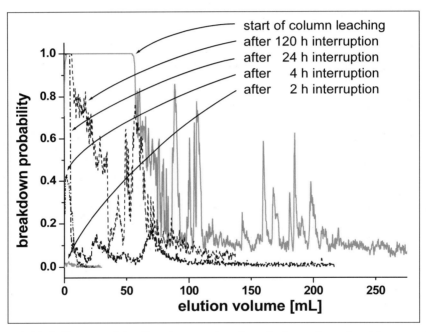

Fig. 9.11. Influence of the flow interruption time during column leaching experiments on the breakdown probability (particle signal) after re-starting the eluent flow.

Given the facts mentioned it seems likely that during the elution break silicate phases dissolve (calcium silicate hydrates sensible to pH-value changes) and subsequently release colloidal fragments. This also happens when the electrolyte concentration is decreased and has been observed by Ferstl (2002) before.

From previous experiments it is known that with regard to colloids the collected 10 mL fractions were not stable (Delay et al. 2006, Wagner et al. 2005). Thus, as soon as possible after completion of the column leaching experiments with directly coupled LIBD (the instrumentation is permanently occupied during flow-through measurements), selected fractions were analyzed by LIBD in order to obtain information on colloid size and concentration. In accordance with the theoretical aspects described above the size decreases in the course of each elution experiment, and so does the concentration. Figs. 9.12-9.14 give an overview. It has to be noted that the collected sample fractions usually do not contain single LIBD peaks and that the obtained results are mean values with regard to the achievable resolution defined by the number of sample fractions per time interval. In the figures this is symbolized by columns of a certain width.

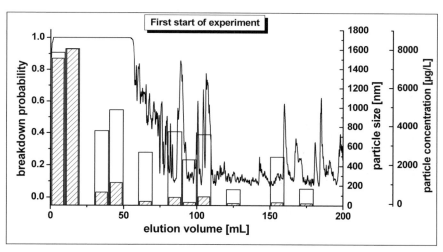

Fig. 9.12. Particles in the effluent of the leaching column after first start of the experiment.

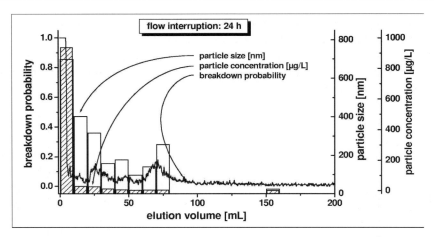

Fig. 9.13. Particles in the effluent of the leaching column after 24 hours of flow interruption.

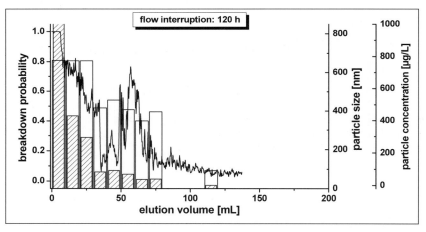

Fig. 9.14. Particles in the effluent of the leaching column after 120 hours of flow interruption.

Right after the start of the column leaching experiments a relatively high release of silicon, calcium, chromium and copper is found (Fig. 9.15). The concentrations decrease in the course of the experiment; the values obtained by ICP-OES measurements include both dissolved and colloidal metals and metal compounds. After a flow interruption of 120 hours the element release follows the same pattern again although the initial concentrations are lower compared to those right after the first start. The reason for the increase in concentrations during flow interruptions is the above mentioned dissolution of solid phases (Ferstl 2002). Since the breakdown probability is increased at the same time, part of the released heavy metals may be bound to colloids. To finally clarify this question further experiments are necessary.

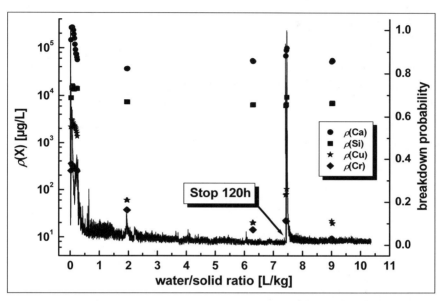

Fig. 9.15. Heavy metal mass concentrations versus water/solid ratio. Both concentration and breakdown probability (particle signal) fall off simultaneously, thus the mobilized heavy metals may be partly colloid bound.

9.4.3 Conclusions

Depending on the experimental conditions, colloids can be mobilized out of slags. The calcium concentration during the elution of the slag is crucial for colloid mobilization; the higher it is, the less mobile the colloids are. The chemical composition of mobilized colloids is inhomogeneous, especially with respect to the elements silicon, calcium and aluminum. Also, form and size vary significantly. The colloids may contain environmentally relevant amounts of heavy metals.

Column leaching experiments can be used to quantify the release of heavy metals and particles out of bottom ashes from municipal waste incinerators. At first, increased concentrations of metal(loid)s are found along with a high count of relatively large particles, both decreases in the course of the experiment. The same behavior is observed after interruptions of the eluent flow; the newly increased values depend on the duration of the flow interruption. With the given experimental setup no significant effects were observed for interruptions of 2 hours or less.

The highly sensitive particle analysis method LIBD has been used to monitor the particle flow *online* by means of a flow-through cell coupled directly to the column leaching unit, and to assign absolute values (polystyrene equivalents) to colloid size and concentration. Both decline in the course of the experiment which is attributed to a decreasing ionic strength and thus widening of the colloids' electrical double layer which in effect opposes agglomeration and leads to smaller particles. The *online* monitoring of the particle flow is most attractive for it was observed that the obtained samples were not stable over time regarding the colloid population.

In nature rain and dry periods often alternate and this can in effect lead to long stagnancy periods of the pore water. As a consequence, toxically relevant concentrations of heavy metals and their compounds can be released temporarily and mobilized both as dissolved and particle borne species.

Acknowledgement

The authors thank the German Federal Ministry of Education and Research for financial support (02WP0516). The authors also thank Birgit Hetzer and Reinhard Sembritzki for their excellent work in the laboratory.

References

Berger W, Kalbe U, Eckardt J, Fischer H, Jansky HJ (2004) Aufbereitung von Referenzmaterialien zur Untersuchung der Eluierbarkeit von Schadstoffen. (Processing of reference materials for examination of contaminant leaching behaviour). Aufbereitungstechnik 45 (11):37-43

Bundschuh T (1999) Entwicklung und Anwendung der Laser-induzierten Breakdown-Detektion zur Quantifizierung aquatischer Kolloide und Actinidenkolloide. Thesis, TU München

Bundschuh T, Hauser W, Kim JI , Knopp R, Scherbaum FJ (2001) Determination of colloid size by 2-D optical detection of laser induced plasma. Coll Surf A 180 (3):285-293

Bundschuh T, Wagner T, Eberhagen I, Hambsch B, Köster R (2005) Detection of biocolloids in aquatic media by nano-particle analyzer (NPA). Spectroscopy - An International Journal 19 (1):69-78

Bundschuh T, Wagner T, Köster R (2005) Laser-induced breakdown detection (LIBD) for the highly sensitive quantification of aquatic colloids. Part I: Principle of LIBD and mathematical model. Particle and Particle Systems Characterization 22:172-180

Chandler AJ, Eighmy TT, Hartlén J, Hjelmar O, Kosson DS, Sawell SE, van der Sloot HA, Vehlow J (1997) Municipal solid waste incinerator residues. Elsevier, Amsterdam

Delay M., Wagner TU, Köster, R, Frimmel, FH (2006) Kopplung einer Säulenelutions-Einheit mit Laser-induzierter Breakdown-Detektion (LIBD) zur empfindlichen Detektion von Kolloiden. GIT 50 (6):552-555

Ferstl W (2002) Physikalisch-chemische Charakterisierung von Kolloiden in Wasser/Reststoff-Systemen: Kolloidgetragene   Schwermetallmobilisierung   in Schlacken. FZKA-Bericht 6736. Forschungszentrum Karlsruhe, Karlsruhe

Johnson CA, Randenburger S, Baccini, P (1995) Acid neutralisation capacity of municipal waste incinerator bottom ash. Environ Sci Technol 29 (1):142-147

Kim JI, Zeh P, Delakowitz B (1992) Chemical interaction of actinide ions with groundwater colloids in gorleben aquifer systems. Radiochim Acta 57/58:147-154

Kitamori T, Yokose K, Suzuki K, Sawada T, Goshi Y (1988) Laser breakdown acoustic effect of ultrafine particle in liquids and its application to particle counting. Jpn J Appl Phys 27:L983-L985

Länderarbeitsgemeinschaft Abfall (LAGA) (1994) LAGA-Merkblatt Entsorgung von Rückständen aus Verbrennungsanlagen für Siedlungsabfälle, Erich Schmidt Verlag, Berlin Bielefeld München

Loveland JP, Bhattacharjee S, Ryan JN, Elimelech M (2003) Colloid transport in a geochemically heterogeneous porous medium: Aquifer tank experiment and modeling. Journal Contaminant Hydrology 65:161-182

McCarthy JF, Zachara JM (1989) Subsurface transport of contaminants. Environ Sci Technol 23 (5):496-502

McDowell-Boyer LM, Hunt JR, Sitar N (1986) Particle transport through porous media. Water Resour Res 22:1901-1921

Radziemski LJ, Cremers DA (1989) Laser-induced plasmas and applications. Marcel Dekker Inc., New York

Ryan JN, Elimelech M (1996) Colloid mobilization and transport in groundwater. Coll Surf A 107:1-56

Scherbaum FJ, Knopp R, Kim, JI (1996) Counting of particles in aqueous solutions by laser induced breakdown photoacoustic detection. Applied Physics B 63 (3):299-306

Stumm W, Morgan JJ (1996) Aquatic chemistry: Chemical equilibria and rates in natural waters. 3rd edn, John Wiley & Sons, New York

Thang NM, Knopp R, Geckeis H, Kim JI, Beck HP (2000) Detection of nanocolloids with flow-field flow fractionation and laser-induced breakdown detection. Anal Chem 72 (1):1-5

Tushar KS, Khilar KC (2006) Review on subsurface colloids and colloid-associated contaminant transport in saturated porous media. Adv colloid interface sci 119 (2-3):71-96

Wagner T, Köster R, Delay M, Frimmel FH (2005) Kopplung einer Säulenelutions-Einheit mit Laser-induzierter Breakdown Detektion (LIBD) zur empfindlichen Detektion von Kolloiden. 71. Jahrestagung der Fachgruppe Wasserchemische Gesellschaft in der GDCh, Bad Mergentheim, 02.-04. Mai, Kurzreferate der Vorträge und Poster, 103-107

Wagner T (2005) Kolloidchemie in aquatischen Systemen – Technische und methodische Weiterentwicklung der Laser-induzierten Breakdown-Detektion (LIBD) von Nano-Teilchen. Thesis, Regensburg

Zevenbergen C, van der Wood T, Bradley, JP, van der Broeck P, Orbons P, van Reeuiwijk LP (1994) Morphological and chemical properties of MSWI bottom ash with respect to the glassy constituents. Has Waste Haz Mater 11 (3):371-383

10 The Role of Colloid Transport in Metal Roof Runoff Treatment

Alexander Schriewer, Konstantinos Athanasiadis, Brigitte Helmreich

Institute of Water Quality Control; Am Coulombwall, 85748 Garching, Germany

10.1 Introduction

Rolled zinc and copper sheets are commonly used as roofing materials and for drainage systems in countries all over the world. The use of copper as a roofing material has a long tradition in Europe back to the 16th century when copper began to be used in Scandinavian countries, originally to prevent buildings from catching fire. Zinc sheet has been used as a roofing material for over two hundred years.

Any metal exposed to atmospheric conditions is subjected to corrosion processes through which corrosion products are formed and accumulate on the surface of the metal roof. During a rain event a part of these products will remain on the surface (patina) and a part will be released and driven in the roof runoff.

Traditionally the roof runoff is sent to sewers through which the rainwater is either directly transported to the receiving water or, in case of combined sewer systems, sent to waste water treatment facilities. On site infiltration of roof runoff as an alternative dewatering concept is under intensive discussion. The use of artificial porous media to eliminate the contaminant load of the roof runoff such as heavy metals, before the water enters the soil and the ground water includes a high risk. One of the most important factors affecting the performance of those infiltration facilities is the contaminant phase distribution, meaning particle, dissolved and colloidal phase.

Mobile colloids in aquatic systems can act as carriers for sorbing contaminants such as heavy metals and therefore may enhance contaminant transport. The colloidal transport through porous media such as aquifers and water supply wells is well documented (Keller and Sirivithayapakorn,

2004; Ryan and Elimelech, 1996; Elimelech et al., 1991; Mc Carthy and Zachara, 1989).

Mc Carthy and Degueldre (1993) have demonstrated that mobile colloids are mostly composed of clay minerals, oxides and hydroxides of Fe and Al, silica, carbonates, and/or natural organic matter (NOM). Manifold different origins for the solid and colloidal fraction in the roof runoff of metal roofs have been reported. According Wallinder et al., (2001) the roof material itself plays an important role on the heavy metal speciation in the roof runoff. By contrast Murakami et al., (2004) demonstrated that the roof orientation has a significant contribution on the origin of the solid and colloidal fraction in the roof runoff.

The aim of this study was to define the phase distribution of the contaminants of an fourteen years old zinc roof runoff, regarding weather conditions, roof orientation and profile of the rain event and how this distribution could affect the performance of three different artificial porous media in respect to contaminant elimination.

10.2 Materials and Methods

10.2.1 Specification of the Local Conditions

The sampling site is located in the campus of the Technical University of Munich in Garching, a rural region of northern Munich. The orographic position of the roof influences the constitution of the primary as well as the secondary deposition, thus the origin of the solid and colloidal fraction in the roof runoff. The highway may be assumed to be a significant source of pollutants. It has traffic sensuous of 130,000 cars per day, 12 % of those are trucks.

The 14 year old zinc roof is divided into eight surfaces, with a sampling surface of 238 m^2. The roof, the gutters and the down spout are made of Ti-Zn. with a maximum titanium percentage by weight of 0.2 % (DIN EN 988, 1996). In order to merge the zinc panels, the standing seam technique was applied. The five chimneys are soldered on the roof with tin-solder which contains fractions of lead. The inclination of all surfaces was 10^0.

10.2.2 Technical System

The roof runoff of 238 m^2 of a zinc roof is channelled into three retention facilities (Fig. 10.1). The difference between these three retention facilities is the artificial porous media used, as a packing material.

The 1st retention facility is packed with a granulated porous concrete coated with iron hydroxide, the 2nd with a natural zeolite named clinoptilolite and the 3rd with a mix of chabazite and philipsite. The monitoring system is composed of a control unit with a digital writer, a flow meter and four automatic samplers.

The control unit equipped with a digital writer controls and triggers the sampling process during a rain event. The signals from the flow meter and from the samplers are digitally recorded (20 s interval modus).

The flow meter device has been industrially calibrated in the range of 0 to 100 L/min and is able to function competently under rain water conditions down to 5 μS/cm.

Every sampler has a capacity of 24 bottles with a volume of 2.5 L each. The sampling process lasts about 2 minutes. During this time the sampler empties away the water which is stagnated in the sampling pipe line from the last process and it sucks a new sample of 250 mL into the collecting bottle. The temperature in the samplers is set to 5 ± 1 ^{0}C.

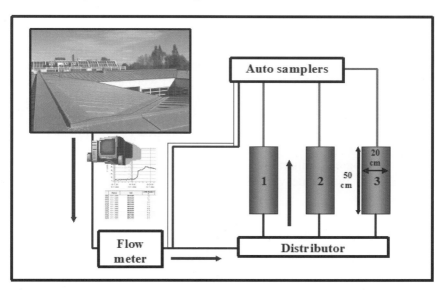

Fig. 10.1. Technical system with monitoring devices

In the case of a precipitation event the roof runoff runs through the flow meter. If the flow is higher than 1 L/min the control unit activates the samplers of the inflow and of the outflow. If the flow becomes then smaller than 0.6 L/min, the control unit stops the sampling process, starting again at 1 L/min. A rain event which generates a runoff flow higher than 1 L/min activates the system again and the new sampling process uses a new bottle of the sampler. The samplers have the potential to arrange a 4x250 mL sampling process in the same bottle. This means that the system can monitor a non stop rain event for more than three hours (192 min).

10.2.3 Sampling

Samples are taken from four different points. (1) From the inflow of the technical system; (2) from the outflow of the retention facility packed with the granulated porous concrete; (3) from the outflow second retention facility packed with clinoptilolite and (4) from the outflow of the third retention facility packed with the mix of chabazite - philipsite. The detailed specification of the packing material is given by Schriewer et al. (2005).

10.2.4 Analytical Methods

Colloid Analysis

Ultrafiltration was performed by means of a pressure filtration unit with stainless-steel filter support (Sartorius). Mixed cellulose ester membrane filters (0.8 μm and 0.025 μm MF-Millipore) and polycarbonate membrane filters (0.8 μm and 0.1 μm Isopore Millipore) were used. The filters were washed with deionised water (resistance 18.2 MΩ, particle concentration below 1μg/L), dried in a desiccator and weighted. After filtration of 300 - 500 mL sample volume the filters were weighted again after drying. The colloid concentration was calculated throughout the difference in weight. The morphological, mineralogical and chemical characterization of the filtered colloids was done via ESEM/EDX and REM/EDX (Jeol 100 CX) measurements.

Cations, pH and Electric Conductivity

Zn^{2+}, Pb^{2+} and Cd^{2+} were determined with atomic absorption spectrometry (Varian SpectrAA-40 and Perkin Elmer AAnalyst 800), calcium and silicium with ICP-OES (Perkin Elmer Optima 4300 DV). The detection limits were 5 μg/L for Pb^{2+}, 0.5 μg/L for Cd^{2+}, 50 μg/L for Zn^{2+}.

The samples have been either filtered sequentially through 0.8 μm and 0.025 μm MF-Millipore or through 0.8 μm and 0.1 μm Isopore membrane filters.

Metal concentrations were measured in the filtered and unfiltered samples. Therefore it was possible to differ between dissolved and solid metals and metals associated with colloids. Concentrations of the residues on the filters were also determined. Sorption of the metals to the filters or the filtration setup was not observed. Additionally, selected filters were analyzed after digestion with aqua regia for heavy metal traces. Electric conductivity measurements were done with a WTW LF 191 conductiviy meter and pH values with a WTW pH 323 pH meter.

Biological Characterisation

Pictures of bacteria colored with DAPI-stain have been pictured with a CLSM 510 mircoscope.

10.3 Results and Discussion

10.3.1 Quality of the Roof Runoff

In the period between May 2004 and May 2005 38 rain events were sampled. All samples were analysed for zinc, pH- value and for electric conductivity. Lead and cadmium were analysed for samples of seventeen rain events. The results obtained are presented in Table 10.1.

Table 10.1. Quality of the zinc roof runoff for the sampled rain events

Runoff	Maximum	Minimum	Median	Average
Zinc (mg/L)	30.0	0.3	5.6	4.3
Lead (µg/L)	31.0	b.d.l	-	-
Cadmium (µg/L)	0.8	b.d.l	-	-
pH	8.4	5.8	6.7	6.7
EC (µS/cm)	242.0	10.0	41.0	50.0

b.d.l.: below detection limit

In the zinc roof runoff the pH value varied between 8.4 and 5.8; the average value was 6.7. Considering that the presence of a zinc panel alters the physico-chemistry of the runoff water a comparison of pH measurements reported in the literature is necessary. Reimann et al., (1997) reported pH levels ranging from 4.0 to 5.0 in rainwater collected in Finland, Norway and Russia with measured SO_4^{2-} concentrations varying between 0.8 and 5 mg/L. Negrel (1998) reported pH values ranging from 4.3 to 6.2 in rainwater samples collected in the Massif Central, in France. Measured SO_4^{2-}concentrations were between 0.4 and 4.9 mg/L. The pH is of major interest at the inflow of the retention facilities due to its impact on the phase distribution of zinc, lead and cadmium.

As expected, different amounts of zinc were washed off the roof surface during all sampled rain events. The concentration of total zinc in the roof runoff varied from 0.3 to 30.0 mg/L and remained even during longer rain events down to a value of about 5 mg/L.

The concentrations of lead in the zinc roof runoff were originally assumed to be rather low because of the use of lead free gasoline technology and the rural location of the field site. Lead could only be detected in some samples at low amounts. In all other rain events the concentration of lead in the roof runoff was lower than 5 µg/L (detection limit). The presence of lead concentration in the roof runoff resulted probably from the sealing material used to seal the chimneys on the roof (Fig. 10.2). The antimony-free solder which was applied on the roof contained 40 % tin and 60 % lead (L-PbSn$_{40}$).

Cadmium was detected only in singular samples with the maximum concentration of 0.8 µg/L.

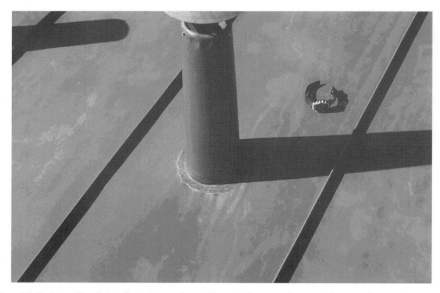

Fig. 10.2. Detail of roof construction: The antimony-free solder sealing material used to seal the chimneys on the roof.

10.3.2 Phase Distribution

With respect to solids, the highest concentrations up to 51.0 mg/L were observed in roof runoff at the beginning of a rain event and were reduced by up to ten times during the remaining rain event (Fig. 10.3). Similar results were reported by Gromaire et al., (1999) in roof runoffs in the city of Paris. This means that a preliminary mechanical treatment of the roof runoff is required for the plane operation of the artificial porous media applied as barrier material in retention facilities for the elimination of heavy metals from roof runoffs.

In respect to the colloidal fraction, a distinct colloidal phase was observed only in some rain events, in spring, and as by the solids the highest colloidal concentrations, till up to 4.4 mg/L, were detected at the beginning of the rain event (Fig. 10.4 and Table 10.2).

These results are indicating that the atmospheric deposition, dry or wet, and not the roofing material itself, were responsible for the high solid concentrations (larger 0.8 μm and colloidal) in the roof runoff at the beginning of a rain event (also see *chemical composition*). Parameters such as dry weather period, wind direction, roof orientation, rain intensity and humidity have also a significant influence on the phase and on the solid concentration in the roof runoff (Wallinder and Leygraf 2001; Wallinder et al., 2000; Cramer and Mc Donald 1990).

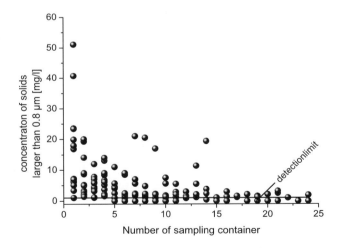

Fig. 10.3. Solid phase distribution in the zinc roof runoff.

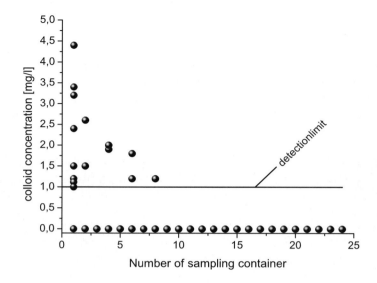

Fig. 10.4. Colloidal phase distribution in the zinc roof runoff.

Table 10.2. Rain events with the maximum colloidal fraction in the zinc roof run-off

Rain event	DWP (h)	Sampling bottles				
		1st	2nd	3rd	4th	5th
05/21/2004	122	3.4	< 1.0	n.a.	< 1.0	< 1.0
05/23/2004	41	4.4	< 1.0	< 1.0	< 1.0	< 1.0
07/18/2004	16	2.4	n.a.	n.a.	1.9	< 1.0
07/20/2004	45	n.a.	1.5	n.a.	2.0	< 1.0
04/18/2005	216	3.2	2.6	< 1.0	< 1.0	< 1.0

n.a.: not analysed (insufficient sample volume); DWP: dry weather period

Characterization of the Colloids

REM/EDX and ESEM/EDX measurements showed that there are not only significant differences of colloid composition between different rain events but also within a single sample of the same rain event. The major differences are size, shape and chemical composition. In each sample different colloids such as silicates, alumosilicates, titanium as well as iron oxides were observed. The use of EDX measurements for quantification of the colloidal composition is inaccurate and it can lead to wrong estimations.

Fig. 10.5 presents an example of a typical ESEM-picture of a 0.1 μm Isopore membrane after the filtration of 50 mL of pre-filtrated (0.8 μm) zinc roof runoff. In spite of the similar appearance of the three marked colloids, the EDX measurements showed a different chemical composition (Fig. 10.6). Colloid A was an iron oxide, B was an alumosilicate and colloid C was a silicate.

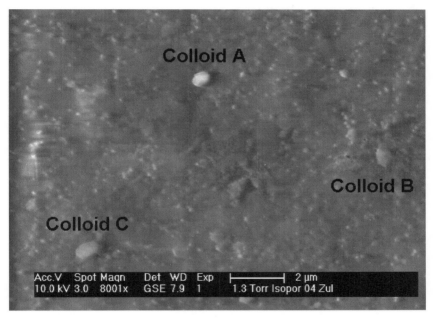

Fig. 10.5. ESEM-picture of 0.1 μm Isopore filter after filtration of 50 mL prefiltrated (0.8 μm) zinc roof runoff

Nonetheless, the highest colloidal concentrations were detected during spring time, no pollen concentration was observed on the 0.1 or 0.025 μm filters. By contrast, high concentrations of pollen were found on the 0.8 μm filters. An additional test of ten minutes ultrasonification of a roof runoff sample with a high pollen concentration was conducted. A few seconds after ultrasonification an instant agglomeration of the pollen concentration could be observed. Rezanejad et al., (2003) also reported that increased particle agglomeration has been observed in samples with polluted pollen.

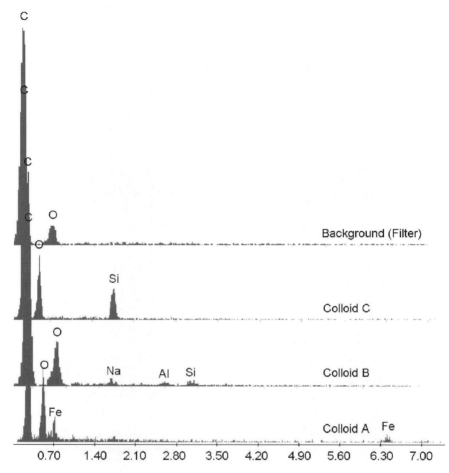

Fig. 10.6. EDX measurements of colloids shown in Fig. 10.5.

Of major importance is the observation that in samples with a high pollen concentration a significant number of bacteria were found. That leads to the conclusion, that bacteria are attached to the pollen. Fig. 10.7(a) shows a 0.1 μm filter of a sample taken in April 2005. Around the silica formation in the centre of the picture a number of bar shaped bacteria could be recognized. A sample from the zinc roof runoff was taken, treated with DAPI-stain and were analysed under confocal laser scanning microscope (CLSM). The results obtained are showed in Fig. 10.7(b). According to Leon-Morales et al. (2004), the colloid – biofilm interactions can influence the colloid bound contaminant transport and the remobilization of the contaminant.

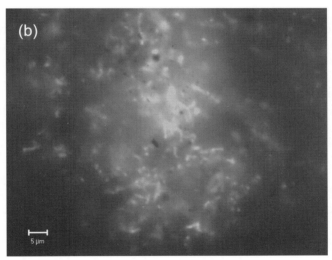

Fig. 10.7. (a) ESEM picture of 0.1 µm Isopore membrane after filtration of 50 mL zinc roof runoff. (b) CLSM image of a sample from the surface of the filter material after a rain event with high pollen concentration.

Metals Associated with Colloids

Zinc associated with colloids could only be observed within the margin of error (Table 10.3). Just as cadmium with amounts of max. 0.8 µg/L, no cadmium associated with colloids could be detected. Agreeing with expectations lead could only be detected as particular matter above 0.8 μm at pH-values near 7.

Table 10.3. Zinc associated with colloids of the rain events shown in table 10.2

| Rain event | DWP (h) | Sampling bottles | | | | |
| | | 1st | 2nd | 3rd | 4th | 5th |
		Concentration of zinc associated with colloids (mg/L)				
05/21/2004	122	1.0	< 1.0	n.a.	< 0.5	< 0.5
05/23/2004	41	0.5	< 0.5	< 0.5	< 0.5	n.a.
07/18/2004	16	< 0.5	n.a.	n.a.	< 0.5	n.a.
07/20/2004	45	< 0.5	< 0.5	n.a.	< 0.5	< 0.5
04/18/2005	216	0.6	< 0.5	< 0.5	< 0.5	< 0.5

n.a. not analysed (insufficient sample volume); DWP: Dry weather period

Considering the appearance of mobile dusts as colloid material, additional analyses of silicium and calcium in the most colloid containing samples were done. The total silicium and calcium concentrations measured ranged from 90 to 340 μg/L and 1.5 to 11.8 mg/L respectively. In the colloidal fraction the calcium and silicium concentrations detected were in the error range. These results are suggesting that the measurable colloidal fractions as a mixture of different single compounds will not facilitate transport of significant amounts of the measured contaminants through treatment facilities or the topsoil.

10.4 Conclusions

The results obtained can be summarised as follows:
- The cover material of the roof and the drainage system were responsible for the high zinc and lead concentrations in the roof runoff
- The phase distribution of the zinc concentration in the roof runoff was dominated by the dissolved phase
- High concentrations of solids (0.8 μm filter) were observed in the roof runoff at the beginning of a rain event. Therefore, a preliminary mechanical treatment of the zinc roof runoff is required for the plane operation of any retention facility applied for the elimination of the heavy metal loading

- A distinct colloidal phase was observed in rain events mainly in spring, and as by the solids the highest colloidal concentrations up to 4.4 mg/L were detected at the beginning of the rain event.
- No dominant colloidal fraction could be defined
- In the case of a zinc roof runoff it was evident that the colloidal transport of zinc was not able to influence the performance of the artificial porous media in respect to zinc elimination.

Acknowledgements

The authors gratefully acknowledge the financial support of this project by the German Research Foundation (Wi 620/13-3).

References

Cramer SD, McDonald LG (1990) in Baboian R, Dean SW (eds.), Corrosion testing and evaluation: Silver Anniversary Volume, ASTM STP 1000. American Society for Testing and Materials, Philadelphia, 241-259

DIN EN 988 (1996) Zink und Zinklegierungen: Anforderungen an gewalzte Flacherzeugnisse für das Bauwesen

Elimelech M (1991) Kinetics of capture of colloidal particles in packed beds under attractive double layer interactions. J Colloid Interface Sci, 146:337-352

Gromaire-Mertz MC, Garnaud S, Gonzalez A, Chebbo G (1999) Characterization of urban runoff pollution in Paris. Water Sci Technol, 39 (2):1-8

He W, Wallinder, I. O., and Leygraf, C. (2001) A laboratory study of copper and zinc runoff during first flush and steady state conditions. Corrosion Sci 43 (1):127-146

Keller AA, Sirivithayapakorn S (2004) Transport of colloids in unsaturated porous media: Explaining large-scale behaviour based on pore-scale mechanisms. Water Resources Research 40

Leon-Morales CF, Leis AP, Strathmann M, Flemming HC (2004) Interactions between laponite and microbial biofilms in porous media: implications for colloid transport and biofilm stability. Water Res 38:3614-3626

McCarthy JF, Degueldre C (1993) Sampling and characterization of colloids and particles in groundwater for studying their role in contaminant transport, 247-315, In: Buffle J, van Leeuwen HP (eds.)

McCarthy JF, Zachara JM (1989) Subsurface transport of contaminants. Environ Sci Technol, 23:496-502

Murakami M, Nakajima F, Furumai H (2004) Modelling of runoff behavior of particle bound polycyclic aromatic hydrocarbons (PAHs) from roads and roofs. Water Res 38 (20):4475-4483

Negrel P, Roy S (1998) Chemistry of rain water in the Massif Central (France): a strontium isotope and major element study. Applied Geochemistry 13 (8):941-952

Reimann CJ, Harrison JJ, Adams JRW (1984) Storm runoff simulation in runoff quality investigations. Gothenburg: Chalmers University of Technology

Rezanejad F, Majd A, Shariatzadeh SMA, Moein M, Aminzadeh M, Mirzaeian M (2003) Effect of air pollution on soluble proteins, structure and cellular material release in pollen of Lagerstroemia indica L Acta Biol Cracov Bot 45 (1):129-132

Ryan JN, Elimelech M (1996) Colloid mobilization and transport in groundwater. Colloids and Surfaces A 107:1-56

Schriewer A, Athanasiadis K, Wilderer PA, Helmreich B (2005) Eliminationsleistung unterschiedlicher Filtermaterialien zum Rückhalt von Zink aus Metalldachabläufen. Jahrestagung der Wasserchemischen Gesellschaft – Fachgruppe in der Gesellschaft Deutscher Chemiker, Bad Mergentheim, ISBN 3-936028-29-X, pp 193-196

Wallinder IO, Verbiest P, He W, Leygraf C (2000) Effects of exposure direction and inclination on the runoff rates of zinc and copper roofs. Corrosion Sci, 42 (8):1471-1487

Index

adsorption isotherms 34, 40, 178, 219

advection-dispersion equation 120, 121, 154

aerobically digested biosolid (ADB) 176, 181, 182, 185, 187, 189, 190, 193-195, 197

aggregation 5, 14, 17, 20, 22, 38, 92, 93, 157, 162, 265

aluminium (Al, Al^{3+}) 224, 226, 274

aminopeptidases 147

anatase 217

arsenic (As) 69, 74, 75, 78-80

atomic force microscopy (AFM) 127, 132

attachment 55-57, 120-125, 128, 130-132, 134-136, 161, 180, 187, 193, 196

attachment (collision) efficiency 57, 122, 123, 126, 137, 156, 160

Bacillus subtilis 145

bacteria 6, 30, 58, 80, 119, 120, 126-128, 131, 146, 148, 149, 152, 153, 155, 256, 277, 282

bacteriophage 119, 124

Basic Stern model 214

biocolloids 6, 119, 120, 121, 129, 131, 135, 155, 158, 254

biodegradation 148

biofilms 80, 143-150, 152, 153, 155, 156, 158, 159, 161-166, 282

biomolecules 131, 132, 134, 135

biosolid colloids 175, 176, 178, 179, 181, 182, 185, 194, 196-198

blocking effects 31, 154

boehmite 217

bottom ashes 251, 252, 259

breakdown probability 254, 255, 265, 266

breakthrough curves (BTC) 15, 19, 20, 42, 46, 48, 64, 65, 71, 72, 73, 75, 124, 129, 130, 156, 157, 159, 179-182, 185, 187, 189-195, 197

Brownian diffusion 127, 128

bypass measurements 37, 47

cadmium (Cd, Cd^{2+}) 103, 105-107, 110, 176, 178, 179, 184-187, 189, 196, 197, 205, 207, 216, 218, 219, 221, 224, 225, 277, 278, 283

calcite colloids 70

calcium (Ca, Ca^{2+}) 12, 14, 16, 31, 106, 113, 148, 159, 160, 162, 163, 218, 228, 261, 262, 264-266, 268, 269, 276, 284

calcium silicate hydrate phases (CSH phases) 252, 264

cation exchange capacity (CEC) 187, 224, 226

CD-MUSIC model 206, 208, 222-224, 228, 235

cerium oxide 218

chabazite 275, 276

charge distribution (CD) model 216, 218, 220

chemical disturbances 7, 8

chemisorption 176, 196

chromium (Cr) 69, 74, 75, 106, 107, 110, 176, 178, 179, 184, 188, 194, 197, 260, 262, 268

clay minerals 30-32, 70, 77, 103, 207, 218, 224, 226, 236, 274

clinoptilolite 275, 276

clogging of pores 4, 5

colloid facilitated transport 4, 5, 19, 21, 22, 24, 30, 37, 48, 49, 153, 251, 252

colloid filtration theory (CFT) 122, 125, 126, 128, 129, 134, 135, 151, 154, 157, 166

colloid mobilization 5, 7, 8, 10-12, 14-16, 18-20, 22-24, 258, 259, 261, 263, 264, 269

Printing: Krips bv, Meppel
Binding: Stürtz, Würzburg